Cadmium Sources and Toxicity

Cadmium Sources and Toxicity

Special Issue Editor
Soisungwan Satarug

MDPI • Basel • Beijing • Wuhan • Barcelona • Belgrade

MDPI

Special Issue Editor
Soisungwan Satarug
The University of Queensland
Australia

Editorial Office
MDPI
St. Alban-Anlage 66
4052 Basel, Switzerland

This is a reprint of articles from the Special Issue published online in the open access journal *Toxics* (ISSN 2305-6304) from 2018 to 2019 (available at: https://www.mdpi.com/journal/toxics/special_issues/Toxicity-Cadmium).

For citation purposes, cite each article independently as indicated on the article page online and as indicated below:

LastName, A.A.; LastName, B.B.; LastName, C.C. Article Title. *Journal Name* **Year**, *Article Number*, Page Range.

ISBN 978-3-03897-984-5 (Pbk)
ISBN 978-3-03897-985-2 (PDF)

Contents

About the Special Issue Editor

Soisungwan Satarug received a B.S. degree in medical technology from Chiang Mai University in Thailand, an M.S. degree in biochemistry from Mahidol University in Thailand, an M.C.H degree in community nutrition from the University of Queensland in Australia, and a Ph.D. degree in biochemistry from the University of Arizona in the United States of America (USA). She received postdoctoral fellowship training in cancer, particularly in the activation of carcinogens by cytochrome P450 enzymes, in the USA (MIT), Japan (NCCRI), Germany (DKFZ), and France (IARC). She was a research scientist at the National Research Centre for Environmental Toxicology in Brisbane, Australia, where she investigated adverse health effects of environmental exposure to toxic metals. Currently, she is a research advisor at the Kidney Disease Research Collaborative, Translational Research Institute, University of Queensland, in Brisbane, Australia. Her research interests revolve around the interplay of nutrition, genetics, and the environment in human disease.

Editorial

Cadmium Sources and Toxicity

Soisungwan Satarug

Kidney Disease Research Collaborative, Faculty of Medicine and Translational Research Institute, The University of Queensland, 37 Kent Street, Woolloongabba, Brisbane 4102, Australia; sj.satarug@yahoo.com.au

Received: 2 May 2019; Accepted: 4 May 2019; Published: 6 May 2019

This special issue of Toxics, Cadmium (Cd) sources and toxicity, consists of one comprehensive review [1], three epidemiologic investigations [2–4] and five laboratory-based investigations [5–9].

A review article highlights environmental exposure to Cd and its association with chronic kidney disease (CKD) together with data from total diet studies (TDS) in which Cd was found to be present in virtually all foodstuffs [1]. Consequently, foods that are frequently consumed in large quantities such as rice, potatoes, wheat, leafy salad vegetables and other cereal crops are the most significant dietary Cd source [1]. Cd levels found in human livers and kidneys are provided together with current standards for tolerable intake, the urinary threshold of Cd and the utility of urinary Cd excretion as a measure of body burden of Cd.

In a cross sectional study of 395 Thai subjects [2], an inverse association was observed between urinary excretion of β_2-microglobulin (β_2MG) and estimated glomerular filtration rate (eGFR) simultaneously with an increase in the prevalence odds of low GFR (eGFR < 60 mL/min/1.73 m^2) in subjects with an elevation of β_2MG excretion, indicative of tubular dysfunction. Thus, a sign of Cd toxicity (tubular dysfunction) was linked to GFR reduction, and an increased risk of CKD, defined as eGFR < 60 mL/min/1.73 m^2. These findings suggest that tubular pathology may have caused nephron atrophy and GFR loss [10].

In a 26-year follow-up study of 7348 residents of the Jinzu River basin in Toyama, a highly polluted area of Japan [3], a 1.49-fold increase in deaths from cancer was observed in women who showed, 26 years earlier, signs of Cd-related kidney pathologies, such as proteinuria and glycosuria. The specific cancer types were uterus, kidney, kidney plus urinary tract. Paradoxically, in men, the risk of lung cancer and the risk of dying from malignant disease were reduced.

In a Chinse cohort study of 429 women, moderate and high environmental Cd exposure levels were associated with an early menarche [4]. Levels of environmental exposure as low, moderate or high were based on rice Cd concentration of 0.07, 0.51 and 3.7 mg/kg, respectively. The age of menopause in three areas did not differ. However, there were 1.3-and 3.7-fold increases in the likelihood of having menarche at age below 13 years in respectively moderate- and high-Cd exposure areas, compared with a low-exposure area. This Chinese finding is consistent with an early onset of puberty seen in Cd-treated female rats [11], thereby suggesting Cd has estrogenic activity.

Effects of inhaled Cd on developing kidneys were examined in Wistar rats [5]. In this study, Cd was administered to pregnant rats from gestation day 8 to day 20 via inhalation of CdCl$_2$ aerosol (17.43 mg/m^3) for 2 hours per day. This procedure delivered a dose of 1.48 mg Cd^{2+}/kg/day. Pregnant rats inhaled normal saline aerosol served as controls. Kidneys from fetuses at gestation day 21 were examined for DNA binding activity of the transcription factor, hypoxia-inducible factor 1 (HIF-1). HIF-1 plays a critical role in the regulation of oxygen consumption, cell survival, growth and development. HIF-1 from kidneys of fetuses of Cd-intoxicated dams showed impairment in DNA-binding activity concomitant with reduced transcript levels for vascular endothelial growth factor (VEGF), one of the HIF-1 regulated genes. However, a compensatory mechanism was apparent as the VEGF protein abundance remained unchanged. These findings suggest potential effects of inhaled Cd on developing kidneys.

Effects of Cd on mature kidneys were examined in male Sprague-Dawley rats [6]. Kidney injury, reflected respectively by 2.2-, 21.7-, and 6.1-fold increases in urinary protein, KIM-1 and β_2MG levels were induced after subcutaneous injections of $CdCl_2$ (0.6 mg/kg) 5 days a week for 12 weeks. Accompanied these urinary indictors of kidney effects were altered expression levels of microRNA (miRNA) in kidney cortex; levels of 44 miRNAs were increased, while levels of another 54 miRNAs were decreased. Thus kidney injury by Cd occurred concurrently with dysregulated miRNA expression in the rat renal cortex. These findings implicated miRNA as mediators of Cd-induced kidney injury.

Effects of Cd on periodontal bone were investigated in male Sprague-Dawley rats, given daily subcutaneous injections of Cd (0.6 mg/kg/day) 5 days a week for 12 weeks [7]. The distance between the cementoenamel junction and the alveolar bone crest was greater in Cd-intoxicated rats than controls. This was taken as evidence for Cd as a possible contributing factor to periodontal disease, thereby explaining an association between elevated body content of Cd and an increased risk of periodontal disease seen in the representative U.S. population.

Effects of Cd on mitochondria were examined in the INS-1 human pancreatic β-cell line [8]. Cd concentration ten-fold below the level causing cell death produced no effects on mitochondrial function, assessed with the energy charge and the synthesis of adenosine triphosphate (ATP). This Cd concentration, however, caused mitochondrial morphological change toward circularity, indicative of fission. The increased circularity suggested mitochondrial adaptive response to low-level Cd. If cellular Cd influx continues, impairment of this organelle may contribute to cellular dysfunction and decreased viability of β-cells, as seen in diabetes.

Therapeutic actions of the anti-diabetic drug, metformin were examined in male Wistar rats given Cd in drinking water (32.5 ppm) alone or Cd plus metformin (200 mg/kg/day) [9]. Cd treatment was found to cause hyperinsulinemia, insulin resistance, adipocyte dysfunction, loss of hepatic insulin sensitivity. Progressive accumulation of triglycerides was also seen in various tissues, while glycogen deposits were diminished in liver, heart, and renal cortex, but was increased in the muscle. Metformin showed a limited therapeutic efficiency on glucose tolerance and lipid accumulation that were induced by Cd.

In summary, this collection of research articles provides an update of knowledge on adverse effects of environmental Cd exposure, such as increased mortality from cancer, especially in women [3], an early menarche onset [4] and an increased risk of chronic kidney disease [2]. Potential effects of inhaled Cd on the development of kidneys in fetuses were evident in a study using pregnant Wistar rats [5]. Work with Sprague-Dawley rats suggested that dysregulation of a range of miRNAs mediated renal Cd toxicity [6] and that Cd contributed to periodontal disease [7]. An early effect of low-dose Cd on mitochondria in human pancreatic β-cells was observed [8]. However, therapeutic efficiency of metformin was not demonstrable when the drug was given to the Wistar rats with Cd-induced metabolic derangements [9].

References

1. Satarug, S. Dietary cadmium intake and its effects on kidneys. *Toxics* **2018**, *6*, 15. [CrossRef] [PubMed]
2. Satarug, S.; Ruangyuttikarn, W.; Nishijo, M.; Ruiz, P. Urinary cadmium threshold to prevent kidney disease development. *Toxics* **2018**, *6*, 26. [CrossRef] [PubMed]
3. Nishijo, M.; Nakagawa, H.; Suwazono, Y.; Nogawa, K.; Sakurai, M.; Ishizaki, M.; Kido, T. Cancer mortality in residents of the cadmium-polluted Jinzu River Basin in Toyama, Japan. *Toxics* **2018**, *6*, 23. [CrossRef] [PubMed]
4. Chen, X.; Zhu, G.; Jin, T. Effects of cadmium exposure on age of menarche and menopause. *Toxics* **2017**, *6*, 6. [CrossRef] [PubMed]
5. Jacobo-Estrada, T.; Cardenas-Gonzalez, M.; Santoyo-Sánchez, M.P.; Thevenod, F.; Barbier, O. Intrauterine exposure to cadmium reduces HIF-1 DNA-binding ability in rat fetal kidneys. *Toxics* **2018**, *6*, 53. [CrossRef] [PubMed]

6. Fay, M.J.; Alt, L.A.C.; Ryba, D.; Salamah, R.; Peach, R.; Papaeliou, A.; Zawadzka, S.; Weiss, A.; Patel, N.; Rahman, A.; et al. Cadmium nephrotoxicity is associated with altered microRNA expression in the rat renal cortex. *Toxics* **2018**, *6*, 16. [CrossRef] [PubMed]

7. Browar, A.W.; Koufos, E.B.; Wei, Y.; Leavitt, L.L.; Prozialeck, W.C.; Edwards, J.R. Cadmium exposure disrupts periodontal bone in experimental animals: Implications for periodontal disease in humans. *Toxics* **2018**, *6*, 32. [CrossRef] [PubMed]

8. Jacquet, A.; Cottet-Rousselle, C.; Arnaud, J.; Julien Saint Amand, K.; Ben Messaoud, R.; Lénon, M.; Demeilliers, C.; Moulis, J.M. The NOAEL Metformin dose is ineffective against metabolic disruption induced by chronic cadmium exposure in Wistar rats. *Toxics* **2018**, *6*, 55. [CrossRef] [PubMed]

9. Sarmiento-Ortega, V.E.; Brambila, E.; Flores-Hernández, J.Á.; Díaz, A.; Peña-Rosas, U.; Moroni-González, D.; Aburto-Luna, V.; Treviño, S. Mitochondrial morphology and function of the pancreatic β-cells INS-1 model upon chronic exposure to sub-lethal cadmium doses. *Toxics* **2018**, *6*, 20. [CrossRef] [PubMed]

10. Schnaper, H.W. The tubulointerstitial pathophysiology of progressive kidney disease. *Adv. Chron. Kidney Dis.* **2017**, *24*, 107–116. [CrossRef] [PubMed]

11. Johnson, M.D.; Kenney, N.; Stoica, A.; Hilakivi-Clarke, L.; Singh, B.; Chepko, G.; Clarke, R.; Sholler, P.F.; Lirio, A.A.; Foss, C.; et al. Cadmium mimics the in vivo effects of estrogen in the uterus and mammary gland. *Nat. Med.* **2003**, *9*, 1081–1084. [CrossRef]

toxics

MDPI

Review

Dietary Cadmium Intake and Its Effects on Kidneys

Soisungwan Satarug

Centre for Kidney Disease Research and Translational Research Institute, The University of Queensland
Diamantina Institute and Centre for Health Services Research, Woolloongabba, Brisbane 4102, Australia;
sj.satarug@yahoo.com.au

Received: 28 February 2018; Accepted: 9 March 2018; Published: 10 March 2018

Abstract: Cadmium (Cd) is a food-chain contaminant that has high rates of soil-to-plant transference. This phenomenon makes dietary Cd intake unavoidable. Although long-term Cd intake impacts many organ systems, the kidney has long been considered to be a critical target of its toxicity. This review addresses how measurements of Cd intake levels and its effects on kidneys have traditionally been made. These measurements underpin the derivation of our current toxicity threshold limit and tolerable intake levels for Cd. The metal transporters that mediate absorption of Cd in the gastrointestinal tract are summarized together with glomerular filtration of Cd and its sequestration by the kidneys. The contribution of age differences, gender, and smoking status to Cd accumulation in lungs, liver, and kidneys are highlighted. The basis for use of urinary Cd excretion to reflect body burden is discussed together with the use of urinary N-acetyl-β-D-glucosaminidase (NAG) and β2-microglobulin (β2-MG) levels to quantify its toxicity. The associations of Cd with the development of chronic kidney disease and hypertension, reduced weight gain, and zinc reabsorption are highlighted. In addition, the review addresses how urinary Cd threshold levels have been derived from human population data and their utility as a warning sign of impending kidney malfunction.

Keywords: β2-microglobulin; body burden indicator; chronic kidney disease; dietary cadmium; exposure assessment; glomerular filtration rate; hypertension; N-acetyl-β-D-glucosaminidase; threshold limit; urine cadmium

1. Introduction

Cadmium (Cd) is a highly persistent environmental toxicant that exhibits higher rates of soil-to-plant transfer than other toxic heavy metals, such as lead (Pb) and mercury (Hg), making Cd a food-chain contaminant of great concern [1,2]. Further, Cd oxide (CdO), which is a highly bioavailable form of Cd, is present in cigarette smoke and polluted air, contributing to elevated Cd concentrations in blood, urine, and tissues of smokers, compared with non-smokers of similar age and gender [3,4]. Historically, consumption of rice contaminated with Cd from zinc mining discharge caused an outbreak of itai-itai disease that affected mostly women [5–7]. The hallmarks of itai-itai disease include severe kidney damage, generalized osteoporosis, osteomalacia, and multiple bone fractures [5–7].

To safeguard population health, safety limits of Cd in the environment and foodstuffs were established [8,9]. A safety limit of 3 mg/kg is applied to soils that are used for producing food crops for human consumption [9], while a 3 µg/L is applied to drinking water [8]. Safety limits, known as maximally permissible concentrations (MPC), have also been established for certain food crops and shellfish that are known as hyper-accumulators of Cd [9]. Currently, the MPC for potatoes is 0.1 mg/kg, while the MPC for rice is 0.4 mg/kg dry grain weight [9]. However, it is argued that MPC for rice should be reduced to 0.2 mg/kg dry grain weight to prevent adverse effects, especially in the populations that consume rice as a dominant energy (calorie) source [10]. This is typical of an Asian diet, which contributes to higher blood and urinary Cd levels in most Asian populations, when compared with other populations [4]. Asian subpopulations have been found to have the highest mean blood Cd among five ethnic groups studied in the U.S. National Health and Nutrition Examination Survey (NHANES), 2011–2012 [11].

In addition, a safe dietary Cd intake guideline and a urinary Cd threshold limit have been established by the Food and Agriculture Organization (FAO) and World Health Organization (WHO) Joint Expert Committee on Food Additives and Contaminants [12,13]. Currently, the FAO/WHO tolerable Cd intake level is 25 μg per kg body weight per month (0.83 μg/kg body weight/day or 58 μg/day for a 70-kg person), while a urinary Cd threshold level is 5.24 μg/g creatinine [14]. A threshold level is defined as a urinary Cd level at which 5% or 10% of the general population shows evidence of an adverse effect. The FAO/WHO tolerable intake level for Cd and the urinary Cd threshold limit were based on lifetime dietary Cd intake limit of 2000 mg per person, and critical kidney Cd concentration of 180–200 μg/g wet kidney weight [12,13].

It is increasingly apparent that adverse kidney effects occur at dietary Cd intake rates that are lower than the FAO/WHO estimated figures [4]. Urinary Cd levels below the threshold limit of 5.24 μg/g creatinine have also been associated with numerous adverse health effects, including chronic kidney disease (CKD) and type-2 diabetes, both of which are increasing in prevalence [4]. Further, cumulative lifetime Cd intake of 1300 mg, not 2000 mg, may increase the risk of developing itai-itai disease [10]. In light of these new data, the FAO/WHO-established safe intake guideline needs to be reassessed, as does the urinary Cd threshold limit.

This review revisits aspects of dietary Cd intake and the effects on kidneys that underpin the FAO/WHO derivation of current threshold limit and tolerable intake levels for Cd. It highlights existing data on levels of Cd accumulation in human lungs, liver and kidneys that vary with age, gender, smoking status, and the presence of diseases. The basis for use of daily urinary Cd excretion rate to reflect total body content of Cd is discussed together with the biomarkers that have been used to quantitate kidney effects of Cd, notably N-acetyl-β-D-glucosaminidase (NAG) and low molecular weight proteins, such as β2-microglobulin (β2-MG). Data on urinary Cd threshold limits derived by the benchmark dose (BMD) method are provided along with their intended use as a warning sign of excessive Cd intake and adverse kidney effects.

2. Cadmium Sources and Intake Estimates

Total diet study (TDS) and food frequency questionnaires (FFQ) have been used to estimate Cd intake rates in μg/day in an average consumer. The TDS is a food safety monitoring program, which is conducted by food authority agencies such as the U.S. Food and Drug Administration (FDA), the Food Standards of Australia and New Zealand (FSANZ), formerly known as the Australia and New Zealand Food Authority (ANZFA), and the European Food Safety Agency (EFSA). It is known also as the "market basket survey" because it involves collection of samples of foodstuffs from supermarkets and retail stores for quantitation of various food additives, pesticide residues, contaminants, and nutrients [14,15]. TDS provides a reasonable method to gauge the relative contribution of each food item to total intake of Cd. As expected, staples that are consumed in large quantities with high frequency contribute the most to total Cd intake. At present, TDS data are available for a limited number of countries, including the United States (U.S.), Australia, Sweden, France, Chile, Spain, Serbia, and Denmark, as reviewed in Satarug et al. [4]. Collectively, TDS data from these countries show that dietary Cd intake levels for the average consumer vary between 8 and 25 μg/day with staples (rice, potatoes, and wheat) forming 40–60% of total dietary Cd intake. Shellfish, crustaceans, mollusks, offal, and spinach are additional Cd sources [4].

In a U.S. study, FFQ estimated a mean dietary Cd intake of 10.4 μg/day (range: 1.74–31.6 μg/day) in women who participated in the Women's Health Initiative [16–18]. In Spain, the mean for dietary intake derived from FFQ was 29.87 μg/day (range: 20.41–41.04 μg/day) for postmenopausal women and 25.29 μg/day (range: 18.62–35 μg/day) for premenopausal women [19,20]. In Japan, the mean Cd intake that was estimated by the FFQ was 26.4 μg/day in one study [21]. In another Japanese study, covering 30 locations nationwide, Cd intake ranged from 12.5 to 70.5 μg/day [22]. The majority of reported dietary Cd intake estimates are within the FAO/WHO tolerable level of 58μg/day for a 70-kg person, with an exception for certain locations in Japan, where intake exceeded the FAO/WHO safe intake guideline [22].

It is widely believed that the TDS method underestimates dietary Cd intake because the distribution of Cd in foods is highly skewed. This skepticism extends to most contaminants that reach foods through unpredictable processes. This problem is the likely cause of a failure to demonstrate an association between estimated Cd intake and the incidence of bone effects and breast cancer [17,19,20,23,24]. In striking contrast, urinary Cd excretion and blood Cd concentration correlate with the risk of developing of many diseases, even if the exposure to Cd is low [4]. A limited utility of TDS and FFQ data has led to an increased use of data from biomonitoring programs (Section 4).

3. An Overview of Cadmium Kinetics

Figure 1 provides an overview of Cd sources, uptake, transport, glomerular filtration, tubular sequestration, and excretion. Cd enters the body through the lungs and gastrointestinal tract in cigarette smoke, polluted air, and food. In cigarette smoke, Cd exists in oxide form (CdO), which is generated as tobacco burns. Cd in plant food crops is mostly in complex with phytochelatin.

Dietary Cd is taken up by the same transporter systems that the body uses to acquire calcium, iron, zinc, and manganese. These transporters may include divalent metal transporter1 (DMT1), Zrt- and Irt-related protein 14 (ZIP14, a member zinc transporter family), the Ca^{2+}-selective channel transient receptor potential vanilloid6 (TRPV6), and human neutrophil gelatinase-associated lipocalin (hNGAL) receptor [25–31]. Cd bound to peptides, small proteins, and phytochelatin may be directly absorbed via transcytosis [30,31]. Cd of dietary origin is transported via the hepatic portal system to the liver, where it induces the synthesis of a metal binding protein, metallothionein (MT), which has a small mass (a molecular weight of 7 kDa) [32–35]. MT contains an unusually high molar content of cysteine indispensable for metal binding [33]. Cd becomes tightly bound to MT, and the complex is denoted as CdMT. Because Cd can exert toxicity as a free ion, CdMT is viewed as a detoxified form. Inhaled Cd induces MT in lungs, and CdMT is formed in situ. CdMT is released into the systemic circulation from enterocytes, liver, and lungs. Because liver cells do not take up the complex [32], CdMT from the gastrointestinal tract may be transported directly to kidneys [36].

In the kidneys, Cd in complexes with proteins, including MT, undergo glomerular filtration and may be taken up by the same receptors and transporter systems in cortical and distal tubules that are involved in reabsorption of proteins and nutrients. These may include ZIP8, ZIP10, ZIP14, DMT1, megalin, hNGAL receptor, TRPV5, and cysteine transporter. Previously, megalin and cubilin were suggested to mediate endocytosis of filtered CdMT [37,38], but this system exhibits only low affinity for CdMT. Thus the megalin and cubilin role in tubular CdMT uptake is questionable. To-date, the mechanisms for tubular CdMT internalization remain unresolved.

Most excreted Cd is believed to have been filtered but not internalized by proximal tubules, because Cd in urine is bound to MT [39]. However, some urinary excretion of CdMT may result from re-entry of exosomes from proximal tubular cells into filtrate [32]. If this phenomenon is incorporated into another parameter, the rate of net tubular sequestration of Cd (TS_{Cd}), then it follows that the filtration rate of Cd (F_{Cd}) equals TS_{Cd} plus the excretion rate (E_{Cd}).

The extremely long half-life of Cd in the human body [40,41] suggests that the majority of Cd that is taken from filtrate is retained indefinitely in tubular cells (a feature of cumulative toxicants). Because the majority of circulating Cd is thought to be bound to albumin, the typical ultrafilterable fraction of $[Cd]_p$ ($[Cd]_{uf}$); consequently, the difference between $[Cd]_{uf}$ and E_{Cd} cannot be determined. Section 4 provides a further discussion on kinetics of Cd and interpretation of human urinary Cd excretion data.

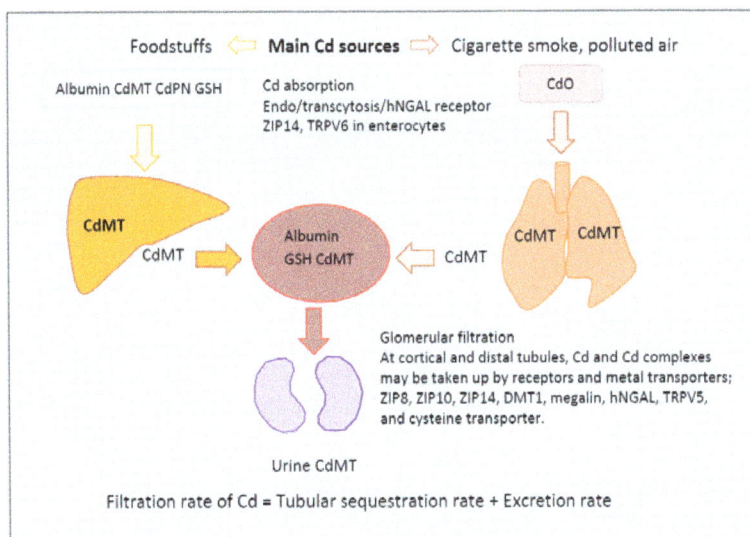

Figure 1. A schematic diagram showing cadmium uptake, transport and urinary excretion. Dietary Cd is absorbed and transported via the hepatic portal system to the liver, where it induces the synthesis of a specific metal binding protein, metallothionein (MT) to which Cd becomes tightly bound. MT-bound Cd is denoted as CdMT. Inhaled Cd induces MT in lungs, and CdMT is formed. CdMT formed by the enterocytes, liver and lungs enters the systemic circulation. Most cells, liver included, do not take up CdMT due to a lack protein internalization mechanism. In the kidneys, Cd, and Cd-complexes, including CdMT undergo glomerular filtration and either excretion or sequestration in proximal tubules. Because Cd in urine is bound to MT, excreted Cd is believed to have been filtered but not taken up by proximal tubules. Some urinary excretion of CdMT may result from re-entry of exosomes from proximal tubular cells into filtrate. CdMT = Metallothionein-bound Cd; CdO = Cadmium oxide; CdPN = Phytochelatin-bound MT; GSH = reduced glutathione; TRPV5 = Transient receptor potential vanilloid6TRPV5; TRPV6 = Transient receptor potential vanilloid6; hNGAL = human neutrophil gelatinase-associated lipocalin; ZIP = Zrt- and Irt-related protein of zinc transporter family; ZIP8 = Zrt- and Irt-related protein 8; ZIP10 = Zrt- and Irt-related protein10; ZIP14 = Zrt- and Irt-related protein 14.

3.1. Gastrointestinal Absorption of Cadmium

Animal and in vitro studies suggest that the absorption of Cd in the gastrointestinal tract is mediated by several transporter systems, which may include divalent metal transporter1, DMT1, Zrt- and Irt-related protein (ZIP) of zinc transporter family, namely ZIP14, and the Ca^{2+}-selective channel, TRPV6 [25–31]. There is also evidence for absorption of dietary Cd by transcytosis mediated by the human neutrophil gelatinase-associated lipocalin (hNGAL) receptor [31]. The divalent metal transporter, DMT1 has the same high affinity for Cd as it does for iron (Km 0.5~1 µM) [25], and was thus thought to be the principal transporter for Cd in the enterocyte [15,16]. However, this transporter can only transport a free Cd ion, while Cd in food and intestinal environment is mostly bound to MT or phytochelatin. Nevertheless there are several potential Cd transporters in enterocytes. The zinc transporter, ZIP14, is highly expressed by the intestinal enterocytes [26,27], as is the Ca^{2+}-selective channel, TRPV6 [28,29]. The calcium binding protein, calbindin may be involved in cytoplasmic transport of Cd, and further research is required to define the transport of Cd to the basolateral cell surface, where it exits the enterocyte into the circulation.

Absorption rates for dietary Cd are influenced by the intake levels and body content of vital metals and elements. Women of reproductive age and children take up more Cd from diet than men because of their low body iron stores and iron deficiency. In a study of 448 healthy, non-smoking Norwegian women (aged 20–55 years, mean 38 years), those who had low body iron stores had 1.42-fold greater blood Cd (0.37 µg/L) than similarly aged women who had normal body iron stores [42]. In the same study, there was an inverse correlation between body iron stores and blood Cd and manganese and the prevalence of high levels of blood Cd and manganese was 26% in those with iron deficiency [42]. A Korean population study reported that women (aged 19–49 years) with iron deficiency had higher mean blood Cd level (1.53 µg/L) than those of the same age and normal body iron status (1.03 µg/L) [43]. Higher dietary zinc intake levels were associated with lower Cd body burden, as assessed by urinary Cd excretion levels [44].

3.2. Glomerular Filtration and Tubular Sequestration of Cadmium

Cd in the systemic circulation is concentrated in erythrocytes, and less than 10% is in the plasma, where it is associated with albumin, amino acids, and glutathione or tightly bound to MT [32].

Protein bound form of Cd is not readily taken up by most cells. Renal tubular cells are an exception because they are equipped for nutrient reabsorption, including virtually all of the proteins in filtrate [45]. In a study that used a microinjection technique, approximately 70–90% of Cd was taken up in the S1-segment of proximal tubules of the rat [46,47]. Uptake of Cd was reduced by a co-injection of zinc or iron [46]. Inhibition of Cd uptake by high concentrations of zinc, iron, and calcium has been demonstrated in another study, using perfused rabbit proximal tubules [47].

The zinc transporters ZIP8, ZIP10, and ZIP14 may mediate the tubular uptake of Cd [48–50]. Transgenic mice with three more copies of the ZIP8 gene accumulated twice as much Cd in the kidney following oral Cd exposure. Elevated ZIP8 expression at the apical membrane of proximal tubular cells accounted for their high sensitivity to Cd toxicity [48]. In mouse kidneys, ZIP8 and ZIP14 at the apical membrane are suggested to mediate the reabsorption of Cd and manganese, especially in the S3 segment of proximal tubules [49]. ZIP10 may also mediate tubular reabsorption of Cd since this zinc transporter is found in high abundance in renal cortical epithelial cells [50].

To-date, the molecular entities mediating the tubular uptake of CdMT have not been resolved (reviewed in [51,52]). Nevertheless, CdMT is taken up and degraded by endosomal and lysosomal protease enzyme systems in tubular cells with consequential release of toxic Cd ions into the cytoplasm. DMT1 was localized to the endosome and the lysosome in rat kidneys, and this suggested that DMT1 might mediate the release of toxic Cd ions [53,54]. This role for DMT1 was later confirmed in an experiment showing that the knockdown of DMT1 expression prevented CdMT-induced toxicity in the proximal tubular cell culture model [55].

The potential for DMT1 in the release of toxic Cd ions also suggests that kidney Cd toxicity may be magnified in iron deficiency state, the conditions in which DMT1 expression levels rise. The localization of FPN1 in the basolateral membrane of proximal tubular cells raises the possibility of involvement of FPN1in mediating Cd efflux. However, the high specificity of FPN1 for iron and cobalt not Cd [56], and only a small fraction of CdMT present at the basolateral membrane suggest that the majority of filtered Cd is retained in tubular cells. This retention may account for the long half-life of Cd in kidneys. The average half-life in kidneys is 14 years. It ranged from 9 to 28 years in a Japanese study [40] and was reported to be 45 years in a modeling study of Swedish kidney transplant donors [41]. The reasons for the large variation in Cd half-life are not apparent.

3.3. Age-, Gender- and Organ-Differentiated Levels of Cadmium Accumulation

In this section, data on measured levels of Cd in human organs are provided in Table 1, which includes data from two Japanese studies [57–64]. One was conducted on residents in an area without Cd contamination [62], and the other was conducted on patients with itai-itai disease and controls [63]. In Table 2 are data on Cd accumulation levels in men and women that include 36 cases of itai-itai disease, and there was only one male case of a total 36 cases [64]. This series exemplifies the preponderance of itai-itai disease in women.

Table 1. Age- and organ-differentiated levels of cadmium accumulation.

Country	Age/Organs	Cadmium (µg/g Wet Tissue Weight)								
Sweden [57]	Age	0–9	10–29	30–39	40–59	60–79	80–99			
	Liver	0.7	0.6	0.6	0.8	1.0	0.6			
	Kidney	2.4	8.8	18.0	19.9	15.0	7.1			
	K/L ratios	3.4:1	15:1	30:1	25:1	15:1	11:1			
Canada I [58]	Age		1–20	21–40	41–60	61–80	81–90			
	Liver		1.0	1.7	2.3	2.2	0.7			
	Kidney		5.4	26.3	41.8	16.4	6.8			
	K/L ratios		5.4:1	16:1	18:1	7.5:1	9.7:1			
Canada II [59]	Age	<10	10–19	20–29	30–39	40–49	50–59	60–69	70–79	>79
	Liver	0.3	0.7	1.4	1.5	1.6	2.2	1.8	1.5	2.5
	Kidney	4.5	5.2	6.8	18.9	41.2	44.2	32.7	23.6	22.8
	K/L ratios	15:1	7.4:1	4.9:1	13:1	26:1	20:1	18:1	16:1	9:1
Australia [60]	Age	2–7	10–19	20–29	30–39	40–49	50–59	60–69	70–79	80–89
	Lung	0.01	0.04	0.22	0.11	0.30	0.14	0.12	0.08	0.03
	Liver	0.21	0.71	0.65	0.95	1.45	0.93	0.94	2.14	1.0
	Kidney	1.63	5.44	9.80	17.8	25.0	22.1	21.6	31.7	8.6
	K/L ratios	7.8:1	7.7:1	15:1	19:1	17:1	24:1	23:1	15:1	8.6:1
Greensland [61]	Age			19–29	30–39	40–49	50–59	60–69	70–79	80–89
	Liver			1.4	2.0	1.7	0.8	1.6	2.6	1.6
	Kidney			12.3	17.8	22.3	18.3	15.8	15.4	5.2
	K/L ratios			8.8:1	8.9:1	13:1	23:1	9.9:1	5.9:1	3.3:1
Japan I [62]	Age	0–1	2–20	21–40	41–60	61–95				
	Liver	0.05	1.1	2.3	1.9	3.6				
	Kidney	0.61	8.4	33.3	69.8	52.3				
	K/L ratios	12:1	7.6:1	15:1	37:1	15:1				
Japan II [63] [a]	Age				46–87	62–97				
	Liver				11.9	69.7				
	Cortex				87.3	36.0				
	Medulla				39.1	25.3				
	K/L ratios				7.3:1	0.5:1				

K/L = Kidney cortex to liver Cd ratio; [a] = Data are from itai-itai disease patients (aged 62–97 years) and controls (aged 46–87 years) [63].

Table 2. Gender differences in levels of cadmium accumulation.

Country	Age/Organs	Cadmium Concentration (µg/g Wet Weight)					
		Males			Females		
		N	Mean	Range	N	Mean	Range
Australia [60]	Age (years)	43	37.05	2–89	18	42.11	3–86
	Lung	43	0.11	0.001–1.15	18	0.17	0.001–1.45
	Liver	43	0.78	0.10–3.23	18	1.36	0.18–3.95
	Kidney	43	14.6	0.72–43.03	18	18.1	1.67–63.25
	Itai-itai disease diagnosis						
Japan III [64]	Age (years)	1	94	-	35	78.5	61–90
	Liver	1	139.0	-	35	62.4	14.4–170.2
	Kidney cortex	1	58.3	-	33	25.6	9.7–112.5
	Kidney medulla	1	36.6	-	32	20.8	8.9–66.7
	Pancreas	1	92.0	-	23	42.8	11.1–102.8
	Thyroid	1	132.1	-	22	35.0	1.9–171.0
	Heart	1	2.9	-	25	0.8	0.2–4.8
	Muscle	1	16.1	-	25	8.5	3.5–14.6
	Aorta	1	3.9	-	24	2.5	0.3–4.7
	Bone	1	2.5	-	25	1.6	0.2–3.8
	Residents of a non-polluted area						
Japan III [64]	Age (years)	36	71.4	60–85	36	72.7	60–91
	Liver	36	7.9	1.3–33.3	36	13.1	3.1–106.4
	Kidney cortex	36	72.1	19.4–200	35	83.9	3.9–252.9
	Kidney medulla	36	18.3	3.5–76.4	35	24.5	4.0–105.0
	Pancreas	7	7.4	3.0–25.9	16	10.5	2.5–29.8
	Thyroid	5	10.6	3.8–35	16	11.9	3.9–56.4
	Heart	7	0.3	0.1–0.5	17	0.4	0.1–1.3
	Muscle	7	1.2	0.3–3.2	16	2.2	0.8–12.4
	Aorta	5	1.0	0.4–2.5	16	1.1	0.3–3.0
	Bone	5	0.4	0.2–0.6	16	0.6	0.2–1.6

3.3.1. Lower Kidney, Higher Liver Cadmium in Itai-Itai Disease Patients

Kidney Cd concentrations in itai-itai disease patients (aged 62–97 years) were dramatically lesser than controls (aged 46–87 years) (Table 1). Kidney Cd concentrations in these patients were 2 times lower than liver Cd levels; the mean of kidney cortex Cd was 36 μg/g wet weight, while the mean of liver Cd was 69.7 μg/g wet weight. The low kidney and high liver Cd in itai-itai disease patients provide strong evidence that diet was the dominant Cd source. Based on Cd content of rice grown in an area, where itai-itai disease was endemic, Cd intake levels were estimated to be over 200 μg/day or 1300 mg over lifetime [10]. The relatively small difference between cortical and medullary Cd content in elderly women with itai-itai disease provide also evidence for their nephron loss at these kidney Cd below a "critical" concentration, discussed below. This is because Cd is reabsorbed primarily by proximal tubules, and cortical Cd content would approach medullary Cd as proximal tubules are lost.

Of note, current Cd risk assessment was based on critical kidney Cd concentration of 180–200 μg/g kidney cortex wet weight [13,14]. However, the mean kidney cortex Cd recorded for itai-itai disease patients, 36 μg/g wet weight (range: 8–133), was far below the critical concentration. This observation casts considerable doubt on the validity of these critical figures [65]. Because of nephron loss, Cd kidney concentrations in people dying from kidney disease were markedly lower than persons dying from other diseases [66].

As shown in Table 2, the mean liver Cd in Australian women was 1.74 fold higher than men [60]. Consistent with Australian data, the mean liver Cd in Japanese women in a low-Cd exposure group was 1.66 fold higher than men. Fractionally, the difference between men and women in kidney cortex content is smaller than the difference in hepatic content. A plausible interpretation is that women have lower iron stores, and adjustments to increase intestinal iron absorption lead to increased absorption and liver uptake of dietary Cd. Redistribution of hepatic Cd to the kidney may be sufficient to cause a higher kidney content of Cd as well, but not so great as to obscure the origin of the increased Cd burden.

3.3.2. Decline in Kidney Cadmium Content in Old Age

Excluding data from itai-itai disease patients, kidney Cd concentrations progressively increased with age, reaching a peak by 40–60 years. Of note, kidney Cd concentrations were consistently lower in the persons older than 60 years, compared to younger age groups. These data could be interpreted to suggest rising Cd exposure in recent times. Most likely, however, these data reflect age-related replacement of tubular cells by fibrosis, which is universal. The peak kidney cortex Cd level was 20, 22, 25, 42, 44, and 70 μg/g kidney cortex wet weight in Sweden, Greenland, Australia, Canada I, Canada II, and Japanese I study series, respectively. The kidney to liver Cd ratio in each corresponding kidney peak group was 25:1, 13:1, 17:1, 18:1, 20:1, and 37:1, respectively. This higher kidney Cd than liver is attributable to a continuing Cd influx (dietary, endogenous reservoirs notably liver, pancreas) to kidneys, as diagrammatically illustrated in Figure 1 and experimentally demonstrated [67,68]. In occupational exposure settings, inhaling relatively high-dose Cd in dust and fumes gave rise to high Cd levels in both liver and kidney (liver Cd 42.3 μg/g wet weight vs. kidney Cd 110 μg/g wet weight) in battery workers [69].

3.3.3. Origin of Cadmium in Kidneys

In the Swedish study, a half of total kidney Cd content (10 μg/g kidney cortex) was estimated to come from food consumption, and the other half was attributed to cigarette smoking [57]. The majority of subjects with high kidney Cd levels (>50 μg/g) were women [57]. In Australian study [60], the mean kidney cortex Cd in high-lung Cd group was 6 μg/g ww higher than the medium-lung Cd group of similar age. The mean kidney Cd in smokers was 5 μg/g ww higher than non-smokers in a large British kidney only study [65]. Further, the mean kidney Cd was 9.7 μg/g ww higher in Australian women with high-lung Cd, when compared to men with similarly high-lung Cd levels although this value did not reach statistical significance. These findings may suggest high pulmonary absorption rates in women, and the redistribution of Cd from lungs to kidneys (Figure 1).

In a study of living kidney transplant donors in Sweden, the rate of kidney Cd accumulation in non-smoker donors was 3.9 µg/g wet weight for every 10-year increase in age [70]. Smoking contributes to an additional 3.7 µg/g wet weight per 10 years. The rate of kidney Cd accumulation in Swedish non-smoker women with low iron stores (serum ferritin ≤ 20 µg/L) was 4.5 µg/g kidney wet weight for every 10-year increase in age.

3.3.4. Urine, Blood and Kidney Cadmium: Data from Kidney Transplant Donors

In an attempt to explore the utility of urine Cd to reflect cumulative lifetime exposure, Akerstrom et al. (2010) analyzed urinary Cd concentrations in relationship to the Cd levels in blood, and kidney biopsies of 109 living kidney transplant donors in Sweden (mean age 51 years, mean kidney Cd 12.9 µg/g wet weight) [71]. A urine-to-kidney Cd ratio of 1:60 was found to correspond to urinary Cd of 0.42 µg/g creatinine and kidney Cd content of 25 µg/g wet weight. In an equivalent analysis using Australian data, a lower urine-to-kidney Cd ratio of 1:20 was assumed because Australian subjects were 10 years younger than the Swedish subjects [72]. A urinary Cd of 1.25 µg/g creatinine corresponded to 25 µg/g Cd/g wet weight, a peak kidney Cd concentration [60]. Section 4 below provides a further discussion on the utility of urinary a quantitative measure of lifetime Cd exposure or intake.

4. Does Urine Cadmium Reflect Total Body Content of Cadmium?

Because TDS and FFQ data are of limited utility in health risk assessment of dietary Cd, there is a paradigm shift to use biomonitoring programs instead of dietary Cd intake estimates (Section 2). In most biomonitoring programs [73–77], single spot urine, and single blood samples are collected for quantitation of various toxicants, which often include ubiquitous toxic heavy metals, namely Cd, Pb, and Hg [73]. Other biologic specimens such as scalp hair and toe nails have sometimes been collected and analysed, but Cd levels in these specimens other than urine have not been rigorously evaluated. Their use remains questionable. Vacchi-Suzzi et al. (2016) have demonstrated good-to-excellent temporal stability of Cd in single spot or first morning void samples, thereby suggesting that urine Cd is suitable for use as a biomarker of long-term Cd exposure in epidemiologic research [78]. An adjustment of spot urine samples for urine creatinine excretion has also been addressed. Adding to the debate on use of urine Cd, this review highlights the fact that urinary Cd excretion can be best used to reflect total body content of Cd. Some certain circumstances that might partially invalidate the assumption for its use are also highlighted.

The utility of urine Cd to reflect total body content of Cd is well founded by Cd levels that accumulated in human organs, notably livers and kidneys, such as those shown in Tables 1 and 2. Kjellstrom and Nordberg (1978) developed the first Cd-toxicokinetics model, using Swedish autopsy data [57,79]. The kinetics model of Cd describes relationships among various parameters that govern the total body content of Cd. These include intake rate from oral and inhalational routes, absorption rate, systemic uptake rate, tissue distribution, half-life, and elimination through bile and urine. The original Kjellstrom and Nordberg model incorporated a single oral absorption rate of 5% for both men and women, and a half-life of Cd in kidneys as 20–30 years. It also assumed that Cd in kidneys comprises one-third of the total body content of Cd, and that 0.005% of the total body content of Cd is excreted in urine per day [79,80]. These assumptions underestimate body burden of Cd. Liver Cd content could be incorporated, given that combined liver and kidney Cd comprises two-thirds of the total body content of Cd.

As more data and knowledge have become available, the Kjellstrom and Nordberg model parameters have now been modified [36,81–85]. The usefulness of modified models has been demonstrated [41,72,86–88]. For instance, modeling of the Cd concentrations in whole blood, plasma, urine, and kidney cortex samples from Swedish kidney transplant donors [41], a calculated daily systemic Cd uptake was 0.0052 µg Cd/kg body weight in men, and 0.0073 µg Cd/kg body weight in women. These systemic uptake rates correspond to an absorption rate between 1.7% and 2.5% in men and 2.4% and 3.5% in women (a 40% higher than men). In another modeling work [72], it is predicted that the dietary intake of Cd at current FAO/WHO tolerable monthly intake rate for 50 years will result in urinary excretion of Cd 0.70–1.85 µg/g creatinine in men and 0.95–3.07 µg/g creatinine in women. These urinary Cd levels have been associated with increased prevalence of CKD in the representative U.S. population (Section 5.3) and other diseases, including liver inflammation, osteoporosis, macular degeneration, hearing loss, depressive symptoms, obesity independent type 2 diabetes, cardiovascular disease, heart disease, breast cancer, and lung cancer in men, reviewed in Satarug et al. [4].

Apparently, urinary Cd excretion is a function of total body content of Cd, nephron numbers, tubular reabsorption capacity, the presence of other diseases, and other conditions, such as hypertension. Urinary Cd excretion rate can thus reflect accurately the total Cd body burden experienced by each individual person. However, interpretation of urinary Cd excretion rates should be done with caution, especially when used to assess Cd body burden in the elderly, people with diabetes, hypertension, and heavy smokers. Because of nephron loss, urinary Cd levels in these subjects can be expected to be lower than similarly age persons, who do not have these conditions. An effect of nephron loss on kidney Cd content is evident from a study that showed persons who died from kidney disease had lower kidney Cd levels than those who died from other diseases [66].

5. Measurement of Effects of Cadmium on Kidneys

Both FAO/WHO, the European Food Safety Agency (EFSA) used kidney tubular effects as the basis for derivation of safe intake levels and urinary Cd threshold limits [13,14,89,90]. Hence, kidney tubular impairment became the most widely studied effect of Cd. This effect is relevant, given that Cd uptake and accumulation in kidney tubular cells is the most extensive and the total amount of Cd in kidneys constitutes one-third of the total body burden (Sections 3.3 and 4). Further, tubular cells contain large number of mitochondria that make them heavily reliant on autophagy to maintain homeostasis and highly susceptible to Cd-induced apoptosis [91–93]. In this section, conventional urinary biomarkers for the assessment of kidney tubular effects are discussed together with urinary Cd threshold levels for these effects. In addition, this section discusses chronic kidney disease (CKD) and other kidney-related effects of Cd that have recently emerged from human population studies.

5.1. Biomarkers for Kidney Effects

A list of conventional urinary biomarkers that researchers have used to investigate tubular effects of Cd is provided in Table 3. These biomarkers are N-acetyl-β-D-glucosaminidase (NAG), lysozyme, total protein and albumin, β2-microglubin (β2-MG), α1-microglobulin (α1-MG), and kidney injury molecule-1 (KIM-1) [94–106]. Urinary levels of these biomarkers were adjusted to urinary creatinine excretion as most studies used single spot or void urine samples. Increased urinary excretion of nutrients, such as glucose, amino acids, calcium, and phosphorus has also been used to reflect tubular effect of Cd [107–109]. As indicated in Table 3, cut-off values of ≥100 mg/g creatinine were used for urinary total protein and ≥30 mg/g creatinine for urinary albumin [98]. These urinary total protein and albumin excretion levels are used also in CKD diagnosis [98]. Cut-off values for other markers, especially NAG, vary widely, depending on study populations and Cd exposure levels (see Section 5.2). Urinary NAG excretion is considered to be proportional to nephron numbers, as these enzymes mostly originate from tubular epithelial cells, and are released upon cell injury [99,107–109]. In a United Kingdom (U.K.) study, a dose–response relationship was observed between urinary Cd and NAG levels [110]. Further, urinary Cd of 0.5 µg/g creatinine was associated with 2.6- and 3.6-fold increases in the prevalence of urinary NAG >2 U/g creatinine, as compared with urinary Cd 0.3 and <0.5 µg/g creatinine, respectively [110].

Urinary α1-MG, β2-MG, and retinol binding protein (RBP) are all low-molecular-weight proteins that have traditionally been used to assess Cd tubular effects [6,99–101]. Another low-molecular-weight protein, namely cystatin C, has recently been evaluated in a rat model for suitability for use in Cd toxicity assessment [111]. Of note, data from Swedish kidney transplant donors suggested that urinary α1-MG excretion was a better marker than RBP or β2-MG, especially in persons with low urinary Cd excretion levels [105]. In the same Swedish donors study, a positive correlation was seen between kidney Cd concentrations and urinary α1-MG levels, while other biomarkers that were measured, such as KIM-1, RBP, and β2-MG did not correlate with kidney Cd concentrations [105]. The mean urinary α1-MG levels in study donors was 7.7 mg/g creatinine (range: 3.25–18.1), and the mean (range) kidney Cd concentrations was 15.0 μg/g wet weight (range: 1.45–55.4) [105].

Table 3. Urinary biomarkers for assessment of kidney effects of cadmium.

Biomarkers	Abnormal Values	Interpretations
NAG	>4 U/g creatinine	Tubular injury, mortality [94–96].
Lysozyme	>4 mg/g creatinine	Tubular injury [97].
Total protein	>100 mg/g creatinine	Glomerular dysfunction, CKD [98].
Albumin	>30 mg/g creatinine	Glomerular dysfunction, CKD [98].
β2-MG	≥1000 μg/g creatinine	Irreversible tubular dysfunction [6,99–101].
β2-MG	≥300 μg/g creatinine	Mild tubular dysfunction, rapid GFR decline [102].
β2-MG	≥145 μg/g creatinine	Increased risk of hypertension, compared with urinary β2MG levels ≤84.5 μg/g creatinine [103].
α1-MG	≥400 μg/g creatinine	Mild tubular dysfunction [104,105]
α1-MG	≥1500 μg/g creatinine	Irreversible tubular dysfunction [6,104].
KIM-1	≥1.6 mg/g creatinine in men, ≥2.4 mg/g creatinine in women	Kidney injury, urinary KIM-1 levels correlate with blood Cd levels [106].

NAG = N-acetyl-β-D-glucosaminidase; β2-MG = β2-microglobulin; α1-MG = α1-microglobulin; KIM-1 = Kidney injury molecule-1.

By virtue of its small mass (MW ~12 kDa), β2-MG is filtered completely, as internalized by the proximal convoluted tubule through megalin-mediated endocytosis, and degraded [112–114]. Approximately 0.3% of filtered β2-MG is excreted in urine. Unlike NAG and lysozyme, β2-MG is produced by most cells in the body. Thus, elevated urinary β2-MG levels may reflect increased systemic β2-MG production or impaired tubular reabsorption [114]. In a sensitivity (renal toxicity) and specificity (non-renal organ toxicity) evaluation in rats, urinary β2-MG detected glomerular injury better than tubular damage [113]. Based on experimental data and clinical outcomes, it is suggested that urinary β2-MG is a predictor of glomerular filtration rate (GFR), and a high urinary β2-MG level could be interpreted to suggest primary glomerular pathologies, leading to protein load and competition of the filtered proteins (β2-MG included) for tubular reabsorption [114].

With respect to Cd effects, urinary β2-MG levels ≥1000 μg/g creatinine are considered to indicate severe and irreversible tubular impairment, while urinary β2-MG levels ≥300 μg/g creatinine are indicative of mild effects. In a prospective study in China, urinary β2-MG levels remained elevated (≥1000 μg/g creatinine) despite a reduction in urinary Cd levels from 11.6 to 9.0 μg/g creatinine over 8-year observation, while urinary albumin excretion recovered [115]. In line with Chinese study, a three-year follow-up study in Korea also suggested the irreversibility of severe tubular impairment in those who had urinary β2-MG levels exceeding 1000 μg/g creatinine [116].

5.2. Urinary Cd Threshold Levels

Currently, the benchmark dose (BMD) method has been used widely to derive a threshold or critical urinary Cd concentration to replace a formerly used no observed adverse effect level (NOAEL) or the lowest observe adverse effect level (LOAEL). Discussion on BMD method can be found in the reports by Crump (1984), Gaylor et al. (1998) and Gainsberg (2012) [117–119]. A threshold for tubular effects is defined as a urinary Cd level at which 5% or 10% of the population shows evidence of abnormal urinary excretion of tubular effect markers. All of the urinary Cd threshold levels shown herein considered a 10% level of risk above background [120–125]. Using data from 790 Swedish women, 53–64 years of age, urinary Cd 0.6–1.1 µg/g creatinine was derived as threshold levels for tubular toxicity [120]. The urinary Cd levels of 0.6–1.2 µg/g creatinine (0.8–1.6 µg/day) in men and 1.2–3.6 µg/g creatinine (0.5–4.7 µg/day) in women were found to be threshold for tubular toxicity, based on data from 828 Japanese subjects (410 men, 418 women), 40–59 years of age, who lived in areas without apparent pollution [121].

Based on data from 547 men to 723 women, aged 50 years or older who were residents of a high-Cd exposure area in Japan, urinary Cd levels of 2.1, 2.6 and 4.1 µg/g creatinine were derived as threshold for abnormal urinary excretion of protein, β2-MG, and NAG in men. The corresponding urinary Cd threshold levels in women were 1.5, 1.4, and 3.1 µg/g creatinine for protein, β2-MG, and NAG, respectively [122]. In this study of residents in a high-Cd exposure area, urinary Cd, β2-MG, and NAG levels were analysed as continuous variables, not being categorized by cut-off values [122]. In a Chinese study, urinary Cd of 0.57–1.84 µg/g creatinine was identified as threshold levels for abnormal urinary β2-MG levels ($\geq$1065 µg/g creatinine), while urinary Cd of 1.19–1.37 µg/g creatinine was identified as threshold levels for abnormal urinary NAG levels ($\geq$5.67 units/g creatinine) [123]. A study of 6103 residents in five high-Cd exposure areas in China, urinary Cd threshold level for a permanent tubular effect (urine β2-MG levels $\geq$ 1000 µg/g creatinine) in men was 2 µg/g creatinine, and 1.69 µg/g creatinine in women [124].

Based on data from occupationally exposed populations in China, urinary Cd threshold levels for abnormal urinary excretion of NAG, β2-MG, and albumin were 2.7, 3.4, and 4.2 µg/g creatinine, respectively [125]. The cut-off values used were 9.8 units/g creatinine, 187.6 µg/g creatinine, and 13.5 mg/g creatinine for NAG, β2-MG, and albumin levels, respectively. These urinary Cd threshold levels in occupationally exposed subjects were slightly higher than in environmentally exposed populations, but all were lower than the FAO/WHO figure. In the same study, a urinary Cd threshold level for abnormally high urinary MT levels ($\geq$388.8 ng/g creatinine) was 3.1 µg/g creatinine [125].

None of the urinary Cd threshold levels that were derived from environmental and occupational exposure situations exceed the FAO/WHO established threshold of $\geq$5.24 µg/g creatinine. Thus, the FAO/WHO figures do not offer health protection. Although Cd has been increasingly associated with disease in tissues and organs other than kidneys [3,4], urinary threshold levels have been derived mostly based on tubular effects. A wide diversity of Cd toxicity levels and toxicity targets requires that urinary Cd threshold levels should be derived for the adverse effects of Cd in many other tissues, such as bone, liver, and retina. In this way, the tissue/organ most sensitive to Cd can be identified, and this organ should be considered as a critical target of Cd toxicity for the derivation of an evidence-based threshold to provide sufficient protection.

5.3. Cadmium and Urine β2-MG: A Revisit

Elevated urinary β2-MG levels that have often been found in people with increased Cd body burden have long been dismissed and have been deemed to not be of clinical relevance. It has further been argued that associations between urinary Cd and commonly measured urinary biomarkers, notably albumin and β2-MG do not reflect toxicity, but reverse causality [126]. In theory, albumin in urine could interfere competitively with CdMT for tubular reabsorption, and thereby increase Cd excretion. Albumin could also impede β2-MG reabsorption. If the patient has a renal disease that is not related to Cd that is rapidly reducing GFR, then Cd excretion would be increased.

However, experimental and clinical outcome data suggest that high urine β2-MG levels could be a result of glomerular pathologies, causing protein load and competition of the filtered proteins for tubular reabsorption [112–114]. Supporting a potential connection between elevated Cd body burden and GFR reduction is an association between higher blood Cd levels and lower eGFR values in adult participants in NHANES 2007–2012 [127]. In addition, a Korean population study has shown that blood Cd levels in the highest tertile were associated with 1.85 mL/min/1.73 m^2 (95% CI: −3.55, −0.16) lower eGFR values, when compared with the lowest tertile [128]. A population-based prospective study in Japan reported that there was a 79% increase in risk of having accelerated GFR decline (10 mL/min/1.73 m^2 over five-year observation period) in those who had urinary β2-MG levels ≥ 300 µg/g creatinine [102]. In a cross-sectional study, urinary β2-MG levels ≥ 145 µg/g creatinine were associated with an increased risk of developing hypertension, as compared with urinary β2-MG levels ≤ 84.5 µg/g creatinine [103].

5.4. Cadmium and Chronic Kidney Disease

Chronic kidney disease (CKD) is a cause of morbidity and mortality, and its prevalence is rising worldwide [129,130]. CKD is defined as an estimated glomerular filtration rate (eGFR) that is below 60 mL/min/1.73 m^2 or urinary albumin to creatinine ratio above 30 mg/g [129,130]. CKD is more prevalent in people with hypertension; the CKD prevalence in 17,794 participants (aged ≥ 20 years) in the U.S. NHANES 1999–2006 was 13.4%, 17.5%, 22%, and 27.5% in those with normal blood pressure, prehypertension, undiagnosed hypertension, and diagnosed hypertension, respectively [131]. CKD prevalence rate in normotensive participants of 13.4% exceeds the 5% acceptable disease prevalence in the general population. In this NHANES 1999–2006 data, urinary Cd levels > 1 µg/L were associated with 41–63% increases in the prevalence odds of CKD and albuminuria [132]. In a separate analysis, blood Cd level of 0.6µg/L or higher showed also an association with risks of developing CKD and albuminuria in NHANES 1999-2006 adult participants [133].

In a recent NHANES 2007–2012 cycle, the overall mean urine Cd level of 0.35 µg/L and mean blood Cd of 0.51 µg/L were lower, when compared with the NHANES 1999–2006 [128]. Such reduction in body burden of Cd in the U.S. population was attributable to a reduction in smoking prevalence, but there was no evidence for a reduction in Cd intake from dietary sources. Despite a reduced population mean urine and blood Cd levels, blood Cd levels > 0.53 µg/L were associated with two-fold increases in prevalence of low GFR (OR 2.21, 95% CI 1.09–4.50) and albuminuria (OR 2.04, 95% CI 1.13–3.69) in an analysis included a subset population (the NHANES 2011–2012) [134]. This was close to the blood Cd level of 0.6 µg/L that was found to be associated with increased risks of developing CKD and albuminuria in adult participants in the NHANES 1999–2006 [133].

Further, an additional increase in risk of albuminuria was seen in Cd-exposed subjects with low zinc status (low serum zinc levels) as OR rose to 3.38 (95% CI, 1.39, 8.28), comparing with those who had higher zinc status [134]. This raises the possibility that CKD results from an increased body burden of Cd. These two conditions have been associated in two NHANES cycles and in cross-sectional studies of other populations, including Korea and China [135,136]. It is also possible that increased urinary Cd, which is the accepted indicator of body burden, may be a consequence of albuminuria or CKD, rather than the cause. Albuminuria may cause also an increase loss of zinc through urine, resulting in trace metal deficiency. Evidence for increased urinary zinc excretion in Cd-exposed subjects in the absence of albuminuria would suggest that Cd can induce urinary loss of zinc whether albumin is present in filtrate or not.

GFR falls if a disease causing albuminuria also destroys glomeruli, or if toxic substances destroy tubular cells after reabsorption from filtrate. Blood pressure rises if GFR falls for any reason, and GFR may fall as a consequence of damage due to hypertension. At a given rate of influx of Cd into plasma from all sources, the plasma Cd concentration is likely to rise as GFR falls. Blood Cd levels ≥ 0.4 µg/L were associated with increased risk of hypertension in Caucasian women (OR 1.54, 95% CI 1.08–2.19), and in Mexican–American women (OR 2.38, 95% CI 1.28–4.40) who participated in the NHANES 1999–2006 [137]. Association between elevated Cd body burden and hypertension development, especially in women, was also seen in Koran and Canadian population studies [138,139]. This would be expected as women are at risk of Cd toxicity due to enhanced Cd uptake (Section 3.1).

Hypertension in Thai women, who were environmentally exposed to Cd, has been associated with increased urinary levels of 20-hydroxyeicosatetraenoic acid (20-HETE), which plays an indispensable role in renal salt balance and blood pressure control [140]. Urinary 20-HETE levels above the median 469 pg/mL were associated with a 90% increase in prevalence odds of hypertension, a four-time increase in odds of having higher urinary Cd levels, and a 53% increase in odds of having higher urinary β2-MG levels [140]. These results link urinary 20-HETE levels to blood pressure increases in Cd-exposed women, thereby providing a plausible mechanism for associated hypertension development.

5.5. Cadmium and Reduced Weight Gain

The body content of Cd assessed by urinary and/or blood Cd levels showed an inverse association with body mass index (BMI), central obesity, and risks of weight gain, and obesity in both children and adults. These have consistently been observed across populations, including the U.S., Belgium, Canada, Korea, and China [77,141–146]. In the U.S. NHANES 1999–2002 participants, an inverse association between body burden (urinary Cd levels) and central obesity was noted [141], while an inverse association between blood Cd and BMI was seen in the NHANES 2003–2010 participants [142]. The Canadian Health Survey 2007–2011 has reported that non-smokers with higher BMI had lower blood and body content of Cd, as reflected by urinary Cd excretion [77]. In a Chinese study, urinary Cd levels that were equivalent to or greater than 2.95 µg/g creatinine were associated with a reduced risk of being overweight [146]. In a Korean study, higher blood Cd levels that were associated with BMI in Korean men (40–70 years) with mean blood Cd of 1.7 µg/L, and mean urinary Cd of 2.13 µg/g creatinine [145].

These observation of lower BMI and lower risk of overweight with higher levels of total body content of Cd are consistent with a reduction in body weight after renal glucose reabsorption is reduced by therapeutically administered sodium glucose cotransporter 2 (SGLT2) inhibitors [147,148]. This was an unexpected outcome from a therapeutic application of glucose reabsorption inhibitors for management of hyperglycaemia or as anti-diabetic drugs [147,148]. This new class of anti-diabetic drugs also show promise in weight reduction and blood pressure control [148,149]. A potential effect of Cd on glucose reabsorption and its contribution to altered body weight homeostasis are discussed below.

Glucose reabsorption in kidneys is mediated by SGLT2, localized to cortical proximal tubular cells, where the bulk of calorie as glucose (an approximate of 160 to 180 gm) is reabsorbed and returned to the systemic circulation daily, under normal physiological conditions [147,148]. An effect of Cd on the activity and/or abundance to SGLT2 in the proximal tubules was deduced from an observation of glycosuria in the subjects with high Cd body burden without hyperglycaemia [94,108]. In a Swedish study, subjects with higher levels of Cd body burden were found to excrete higher levels of citrate, 3-hydroxyisovalerate, and 4-deoxy-erythronic acid, which are the biomarkers of mitochondria [150]. In this study, a positive association was seen also between urinary Cd and 8-oxo-deoxyguanosine, a marker of increased systemic oxidative stress. Increased urinary citrate levels may be secondary to Cd effect on tubular reabsorption of filtered citrate rather than the spill out of citrate and other organic anions due to mitochondrial damage. This is because no correlation was seen between urinary citrate and NAG levels.

Interestingly, a study of 168 Thai subjects in a high-Cd exposure area and 100 controls also observed increases in urinary citrate levels with Cd, and the authors suggested that Cd may have a direct effect on mitochondrial citrate metabolism because of the strength of an association between urinary Cd and urinary citrate levels persisted after adjustment for age, smoking status, and severity of tubular impairment, assessed by urine β2-MG levels [151]. Support of these Swedish and Thai data included an effect of Cd on mitochondrial oxidative phosphorylation in tubular cells, causing a reduction in ATP output [91], and a fall in the abundance of the Na/K-ATPase and its sodium transport activity in tubular epithelial cells that were treated with Cd [52,152]

5.6. Cadmium and Depressed Serum Zinc: Role for Impaired Zinc Reabsorption

Zinc is the second most abundant metal in the body, and the contribution of kidneys to zinc homeostasis is well established [153]. However, there are limited studies that assessed Cd effects on zinc homeostasis and tubular zinc reabsorption. A potential Cd effect on zinc homeostasis comes from the Belgian study, including 959 men and 1018 women, aged 20–80 years, which observed depressed serum zinc levels in those with elevated Cd body burden [154]. The depressed serum zinc levels persisted even after subjects with occupational exposure to metals were excluded [154]. Likewise, reduced serum zinc levels were found to be associated with higher blood Cd levels in another study of 299 healthy Croatian men, 20–55 years of age [155]. Of note, women in the Belgian study were found to have lower serum zinc (mean 12.6 μM, range 6.3–23.2 μM) than men (mean 13.1 μM, range 6.5–23.0 μM). A Thai study observed also lower mean serum zinc in women (18.4 μM), as compared to men (21 μM). In this Thai study, lower fractional zinc reabsorption levels were associated with higher Cd body burden, higher serum copper to zinc ratios, and higher tubular impairment levels, as assessed by the urinary excretion of β2-MG [156].

In an Australian autopsy study, levels of zinc in liver and kidney cortex decrease with rising age and Cd levels [157]. In this study, liver and kidney MT levels were not quantified. However, a regression model analysis showed that a large fraction of zinc in kidney cortex was associated with the MT pool [157]. In the liver, however, there was much less zinc in MT than in non-MT pool. These findings are consistent with MT metamorphism, which explains the different zinc and copper contents in MT molecules from different tissues and organs [33]. Australian data suggested that zinc in kidney was mostly bound to MT, whereas the majority of zinc in liver was not associated with MT.

By immunohistochemistry, in human kidneys cortex MT-1/2 was found mostly in the cytoplasm and nuclei of proximal tubular cells, and to a less extent in distal tubules, but not in the glomeruli, or associated interstitial and vascular elements [158]. MT-1/2 was found in proximal and distal tubules of rat kidneys [152]. Another form of MT (MT-3) was found to be expressed in high abundance in human kidneys, especially in the distal tubule [159]. However, MT-3 is known to bind Cd less vividly than MT-1/2. In primary culture of human proximal tubular cells, MT-1/2 was induced by as little as 0.5 μM Cd [158]. A study in Thai subjects in high-Cd exposure area reported that MT transcript levels in leucocytes increased with increasing blood and urinary Cd levels [160]. Further, the high levels of MT transcript in leucocytes were associated with reduced urinary albumin and β2-MG levels, suggesting a reduction in Cd toxicity as MT levels increased [160].

The influence of MT and Cd concentrations on kidney zinc and copper concentrations is suggested by data from two additional Turkish population studies [161,162]. In one study, the AG and GG variants of the MT2A gene promoter were associated with higher kidney cortex Cd levels than the AA variant, but there were no differences in zinc or copper levels [161]. In the other study, the GG variant was associated with higher blood Cd levels, but lower blood Zn levels, when compared with the AA and AG variants [162]. In contrast, data from a large Japanese (Nagoya) study (749 men and 2025 women, aged 39–75 years) observed no differences in serum MT, Cd, or zinc levels across the three MT2A promoter variants (AA, GA, GG). Of interest, the GG variant was associated with an increased risk of developing CKD or diabetes in Japanese subjects [163]. Collectively, these findings suggest a closer link between Cd, MT, and zinc homeostasis than copper and further investigation is required to dissect the link that may exist between MT, Cd, zinc, and the development of CKD and diabetes.

6. Conclusions

Currently, dietary Cd intake is estimated to be between 8 and 25 μg/day in various populations. These are within the FAO/WHO tolerable intake level of 58 μg/day for a 70-kg person. Kidney cortical Cd concentrations increase progressively with age, reaching a peak by 40–60 years. The recorded peak kidney cortical Cd accumulation of 20–70 μg/g wet weight is also well below critical kidney Cd concentrations of 180–200 μg/g kidney. However, population research data reviewed herein suggest that Cd has adverse effects on kidneys at Cd intake rates and kidney Cd concentrations that are lower than these estimated figures. Elevated urinary excretion of a low-molecular weight protein β2-MG and NAG, termed tubular proteinuria and enzymenuria, have been used to reflect kidney toxic effects of Cd ever since the discovery of β2-MG in urine of Cd-exposed humans. Supporting an effect of Cd on kidneys is an association between GFR reduction and increased urinary β2-MG levels.

Other possible kidney effects of Cd may include an inhibition of glucose reabsorption, and reduced zinc reabsorption by the kidneys, thereby affecting energy and zinc homeostasis. These adverse effects of Cd on the kidneys have been observed at urinary Cd levels below an established threshold limit of urinary Cd of 5.24 μg/g creatinine. These observations cast considerable doubt on the validity of current "tolerable" intake level for Cd and its "critical" kidney concentrations. There is an urgent need to reassess the Cd toxicity threshold limit, as it currently does not afford the protection that it should to prevent excessive Cd exposure and its adverse effects. Public health measures are needed to minimize Cd contamination of the food-chain. Risk reduction measures are also required to reduce air pollution, smoking, workplace exposure, and gastrointestinal absorption of Cd, especially for populations that are deemed to be of increased risk of exposure from all sources.

Acknowledgments: The author thanks Kenneth Phelps for advice and insightful comments on renal physiology fundamental to understanding renal handling of cadmium. The author thanks Frank Thevenod for comments and sharing knowledge on cadmium transporters. The author thankfully acknowledges Shigeki Shibahara, Kazumichi Furuyama and Shigeru Taketani for the support given to author's research in Japan. Support and encouragement from CKDR colleagues (Glenda Gobe, David Vesey and David Johnson) are gratefully acknowledged.

Conflicts of Interest: The authors declare no conflict of interest.

References

1. ATSDR (Agency for Toxic Substances and Disease Registry). *Toxicological Profile for Cadmium*; Department of Health and Humans Services, Public Health Service, Centers for Disease Control and Prevention: Atlanta, GA, USA, 2012.

2. McLaughlin, M.J.; Singh, B.R. Cadmium in soils and plants. In *Developments in Plant and Soil Sciences*; McLaughlin, M.J., Singh, B.R., Eds.; Kluwer Academic Publishers: Dorddrecht, The Netherlands; Boston, London, 1999; Volume 85, pp. 1–7.

3. Satarug, S.; Moore, M.R. Adverse health effects of chronic exposure to low-level cadmium in foodstuffs and cigarette smoke. *Environ. Health Perspect.* **2004**, *112*, 1099–1103. [CrossRef] [PubMed]

4. Satarug, S.; Vesey, D.A.; Gobe, G.C. Current health risk assessment practice for dietary cadmium: Data from different countries. *Food Chem. Toxicol.* **2017**, *106*, 430–445. [CrossRef] [PubMed]

5. Aoshima, K. Epidemiology of renal tubular dysfunction in the inhabitants of a cadmium-polluted area in the Jinzu river basin in Toyama Prefecture. *Tohoku J. Exp. Med.* **1987**, *152*, 151–172. [CrossRef] [PubMed]

6. Horiguchi, H.; Aoshima, K.; Oguma, E.; Sasaki, S.; Miyamoto, K.; Hosoi, Y.; Katoh, T.; Kayama, F. Latest status of cadmium accumulation and its effects on kidneys, bone, and erythropoiesis in inhabitants of the formerly cadmium-polluted Jinzu River Basin in Toyama, Japan, after restoration of rice paddies. *Int. Arch. Occup. Environ. Health* **2010**, *83*, 953–970. [CrossRef] [PubMed]

7. Baba, H.; Tsuneyama, K.; Kumada, T.; Aoshima, K.; Imura, J. Histological analysis for osteomalacia and tubulopathy in itai-itai disease. *J. Toxicol. Sci.* **2014**, *39*, 91–96. [CrossRef] [PubMed]

8. WHO. *IPCS (International Programme on Chemical Safety) Environmental Health Criteria 134: Cadmium*; WHO: Geneva, Switzerland, 1992.

9. Codex Alimentarius Commission. (2015). CODEX STAN 193–1995. In *General Standard for Contaminants and Toxins in Food and Feed*; Food and Agriculture Organization of the United Nations and World Health Organization: Rome, Italy, 2015.

10. Kubo, K.; Nogawa, K.; Kido, T.; Nishijo, M.; Nakagawa, H.; Suwazono, Y. Estimation of benchmark dose of lifetime cadmium intake for adverse renal effects using hybrid approach in inhabitants of an environmentally exposed river basin in Japan. *Risk Anal.* **2017**, *37*, 20–26. [CrossRef] [PubMed]

11. Awata, H.; Linder, S.; Mitchell, L.E.; Delclos, G.L. Biomarker levels of toxic metals among Asian populations in the United States: NHANES 2011–2012. *Environ. Health Perspect.* **2017**, *125*, 306–313. [CrossRef] [PubMed]

12. Food and Agriculture Organization of the United Nations (FAO); World Health Organization (WHO). *Evaluation of Certain Food Additives and Contaminants: Forty-First Report of the Joint FAO/WHO Expert Committee on Food Additives*; WHO Technical Report Series No. 837; WHO: Geneva, Switzerland, 1993.

13. Food and Agriculture Organization of the United Nations (FAO); World Health Organization (WHO). Summary and Conclusions. In Proceedings of the Joint FAO/WHO Expert Committee on Food Additives Seventy-Third Meeting, Geneva, Switzerland, 8–17 June 2010.

14. Callan, A.; Hinwood, A.; Devine, A. Metals in commonly eaten groceries in Western Australia: A market basket survey and dietary assessment. *Food Addit. Contam. A* **2014**, *31*, 1968–1981. [CrossRef] [PubMed]

15. Calafat, A.M. The U.S. National Health and Nutrition Examination Survey and human exposure to environmental chemicals. *Int. J. Hyg. Environ. Health* **2012**, *215*, 99–101. [CrossRef] [PubMed]

16. Awata, H.; Linder, S.; Mitchell, L.E.; Delclos, G.L. Association of dietary intake and biomarker levels of arsenic, cadmium, lead, and mercury among Asian populations in the U.S.: NHANES 2011–2012. *Environ. Health Perspect.* **2017**, *125*, 314–323. [PubMed]

17. Adams, S.V.; Quraishi, S.M.; Shafer, M.M.; Passarelli, M.N.; Freney, E.P.; Chlebowski, R.T.; Luo, J.; Meliker, J.R.; Mu, L.; Neuhouser, M.L.; et al. Dietary cadmium exposure and risk of breast, endometrial, and ovarian cancer in the Women's Health Initiative. *Environ. Health Perspect.* **2014**, *122*, 594–600. [CrossRef] [PubMed]

18. Quraishi, S.M.; Adams, S.M.; Meliker, J.R.; Li, W.; Luo, J.; Neuhouser, M.L.; Newcomb, P.A. Urinary cadmium and estimated dietary cadmium in the Women's Health Initiative. *J. Expo. Sci. Environ. Epidemiol.* **2016**, *26*, 303–308. [CrossRef] [PubMed]

19. Puerto-Parejo, L.M.; Aliaga, I.; Canal-Macias, M.L.; Leal-Hernandez, O.; Roncero-Martín, R.; Rico-Martín, S.; Moran, J.M. Evaluation of the dietary intake of cadmium, lead and mercury and its relationship with bone health among postmenopausal women in Spain. *Int. J. Environ. Res. Public Health* **2017**, *14*, 564. [CrossRef] [PubMed]

20. Lavado-García, J.M.; Puerto-Parejo, L.M.; Roncero-Martín, R.; Moran, J.M.; Pedrera-Zamorano, J.D.; Aliaga, I.J.; Leal-Hernández, O.; Canal-Macias, M.L. Dietary intake of cadmium, lead and mercury and its association with bone health in healthy premenopausal women. *Int. J. Environ. Res. Public Health* **2017**, *14*, 1437. [CrossRef] [PubMed]

21. Itoh, H.; Iwasaki, M.; Sawada, N.; Takachi, R.; Kasuga, Y.; Yokoyama, S.; Onuma, H.; Nishimura, H.; Kusama, R.; Yokoyama, K.; et al. Dietary cadmium intake and breast cancer risk in Japanese women: A case-control study. *Int. J. Hyg. Environ. Health* **2014**, *217*, 70–77. [CrossRef] [PubMed]

22. Ikeda, M.; Nakatsuka, H.; Watanabe, T.; Shimbo, S. Estimation of daily cadmium intake from cadmium in blood or cadmium in urine. *Environ. Health Prev. Med.* **2015**, *20*, 455–459. [CrossRef] [PubMed]

23. Larsson, S.C.; Orsini, N.; Wolk, A. Urinary cadmium concentration and risk of breast cancer: A systematic review and dose-response meta-analysis. *Am. J. Epidemiol.* **2015**, *182*, 375–380. [CrossRef] [PubMed]

24. Lin, J.; Zhang, F.; Lei, Y. Dietary intake and urinary level of cadmium and breast cancer risk: A meta-analysis. *Cancer Epidemiol.* **2016**, *42*, 101–107. [CrossRef] [PubMed]

25. Garrick, M.D.; Dolan, K.G.; Horbinski, C.; Ghio, A.J.; Higgins, D.; Porubcin, M.; Moore, E.G.; Hainsworth, L.N.; Umbreit, J.N.; Conrad, M.E.; et al. DMT1: A mammalian transporter for multiple metals. *Biometals* **2003**, *16*, 41–54. [CrossRef]

26. Fujishiro, H.; Hamao, S.; Tanaka, R.; Kambe, T.; Himeno, S. Concentration-dependent roles of DMT1 and ZIP14 in cadmium absorption in Caco-2 cells. *J. Toxicol. Sci.* **2017**, *42*, 559–567. [CrossRef] [PubMed]

27. Jorge-Nebert, L.F.; Gálvez-Peralta, M.; Landero Figueroa, J.; Somarathna, M.; Hojyo, S.; Fukada, T.; Nebert, D.W. Comparing gene expression during cadmium uptake and distribution: Untreated versus oral Cd-treated wild-type and ZIP14 knockout mice. *Toxicol. Sci.* **2015**, *143*, 26–35. [CrossRef] [PubMed]

28. Kovacs, G.; Danko, T.; Bergeron, M.J.; Balazs, B.; Suzuki, Y.; Zsembery, A.; Hediger, M.A. Heavy metal cations permeate the TRPV6 epithelial cation channel. *Cell Calcium* **2011**, *49*, 43–55. [CrossRef] [PubMed]

29. Kovacs, G.; Montalbetti, N.; Franz, M.C.; Graeter, S.; Simonin, A.; Hediger, M.A. Human TRPV5 and TRPV6: Key players in cadmium and zinc toxicity. *Cell Calcium.* **2013**, *54*, 276–286. [CrossRef] [PubMed]

30. Fujita, Y.; el Belbasi, H.I.; Min, K.S.; Onosaka, S.; Okada, Y.; Matsumoto, Y.; Mutoh, N.; Tanaka, K. Fate of cadmium bound to phytochelatin in rats. *Res. Commun. Chem. Pathol. Pharmacol.* **1993**, *82*, 357–365. [PubMed]

31. Langelueddecke, C.; Lee, W.K.; Thévenod, F. Differential transcytosis and toxicity of the hNGAL receptor ligands cadmium-metallothionein and cadmium-phytochelatin in colon-like Caco-2 cells: Implications for in vivo cadmium toxicity. *Toxicol. Lett.* **2014**, *226*, 228–235. [CrossRef] [PubMed]

32. Sabolić, I.; Breljak, D.; Skarica, M.; Herak-Kramberger, C.M. Role of metallothionein in cadmium traffic and toxicity in kidneys and other mammalian organs. *Biometals* **2010**, *23*, 897–926. [CrossRef] [PubMed]

33. Krężel, A.; Maret, W. The functions of metamorphic metallothioneins in zinc and copper metabolism. *Int. J. Mol. Sci.* **2017**, *18*, 1237. [CrossRef] [PubMed]

34. Prozialeck, W.C.; Edwards, J.R. Early biomarkers of cadmium exposure and nephrotoxicity. *Biometals* **2010**, *23*, 793–809. [CrossRef] [PubMed]

35. Prozialeck, W.C.; Edwards, J.R. Mechanisms of cadmium-induced proximal tubule injury: New insights with implications for biomonitoring and therapeutic interventions. *J. Pharmacol. Exp. Ther.* **2012**, *343*, 2–12. [CrossRef] [PubMed]

36. Elsenhans, B.; Strugala, G.J.; Schafer, S.G. Small-intestinal absorption of cadmium and the significance of mucosal metallothionein. *Hum. Exp. Toxicol.* **1997**, *16*, 429–434. [CrossRef] [PubMed]

37. Klassen, R.B.; Crenshaw, K.; Kozyraki, R.; Verroust, P.J.; Tio, L.; Atrian, S.; Allen, P.L.; Hammond, T.G. Megalin mediates renal uptake of heavy metal metallothionein complexes. *Am. J. Physiol. Renal Physiol.* **2004**, *287*, F393–F403. [CrossRef] [PubMed]

38. Wolff, N.A.; Abouhamed, M.; Verroust, P.J.; Thevenod, F. Megalin-dependent internalization of cadmium-metallothionein and cytotoxicity in cultured renal proximal tubule cells. *J. Pharmacol. Exp. Ther.* **2006**, *318*, 782–791. [CrossRef] [PubMed]

39. Wolf, C.; Strenziok, R.; Kyriakopoulos, A. Elevated metallothionein-bound cadmium concentrations in urine from bladder carcinoma patients, investigated by size exclusion chromatography-inductively coupled plasma mass spectrometry. *Anal. Chim. Acta* **2009**, *631*, 218–222. [CrossRef] [PubMed]

40. Suwazono, Y.; Kido, T.; Nakagawa, H.; Nishijo, M.; Honda, R.; Kobayashi, E.; Dochi, M.; Nogawa, K. Biological half-life of cadmium in the urine in the habitats after cessation of exposure. *Biomarkers* **2009**, *14*, 77–81. [CrossRef] [PubMed]

41. Fransson, M.N.; Barregard, L.; Sallsten, G.; Akerstrom, M.; Johanson, G. Physiologically-based toxicokinetic model for cadmium using Markov-Chain Monte Carlo analysis of concentrations in blood, urine, and kidney cortex from living kidney donors. *Toxiol. Sci.* **2014**, *141*, 365–376. [CrossRef] [PubMed]

42. Meltzer, H.M.; Brantsaeter, A.L.; Borch-Iohnsen, B.; Ellingsen, D.G.; Alexander, J.; Thomassen, Y.; Stigum, H.; Ydersbond, T.A. Low iron stores are related to higher blood concentrations of manganese, cobalt and cadmium in non-smoking, Norwegian women in the HUNT 2 study. *Environ. Res.* **2010**, *110*, 497–504. [CrossRef] [PubMed]

43. Suh, Y.J.; Lee, J.E.; Lee, D.H.; Yi, H.G.; Lee, M.H.; Kim, C.S.; Nah, J.W.; Kim, S.K. Prevalence and relationships of iron deficiency anemia with blood cadmium and vitamin D levels in Korean women. *J. Korean Med. Sci.* **2016**, *31*, 25–32. [CrossRef] [PubMed]

44. Vance, T.M.; Chun, O.K. Zinc intake is associated with lower cadmium burden in U.S. adults. *J. Nutr.* **2015**, *145*, 2741–2748. [CrossRef] [PubMed]

45. Nielsen, R.; Christensen, E.I.; Birn, H. Megalin and cubilin in proximal tubule protein reabsorption: From experimental models to human disease. *Kidney Int.* **2016**, *89*, 58–67. [CrossRef] [PubMed]

46. Barbier, O.; Jacquillet, G.; Tauc, M.; Poujeol, P.; Cougnon, M. Acute study of interaction among cadmium, and zinc transport along the rat nephron in vivo. *Am. J. Physiol. Ren. Physiol.* **2004**, *287*, F1067–F1075. [CrossRef] [PubMed]

47. Wang, Y.; Zalups, R.K.; Barfuss, D.W. Potential mechanisms involved in the absorptive transport of cadmium in isolated perfused rabbit renal proximal tubules. *Toxicol. Lett.* **2010**, *193*, 61–68. [CrossRef] [PubMed]

48. Schneider, S.N.; Liu, Z.; Wang, B.; Miller, M.L.; Afton, S.E.; Soleimani, M.; Nebert, D.W. Oral cadmium in mice carrying 5 versus 2 copies of the *Slc39a8* gene: Comparison of uptake, distribution, metal content, and toxicity. *Int. J. Toxicol.* **2014**, *33*, 14–20. [CrossRef] [PubMed]

49. Fujishiro, H.; Yano, Y.; Takada, Y.; Tanihara, M.; Himeno, S. Roles of ZIP8, ZIP14, and DMT1 in transport of cadmium and manganese in mouse kidney proximal tubule cells. *Metallomics* **2012**, *4*, 700–708. [CrossRef] [PubMed]

50. Kaler, P.; Prasad, R. Molecular cloning and functional characterization of novel zinc transporter Zip10 (Slc39a10) involved in zinc uptake across rat renal brush-border membrane. *Am. J. Physiol. Ren. Physiol.* **2007**, *292*, F217–F229. [CrossRef] [PubMed]

51. Thevenod, F.; Wolff, N.A. Iron transport in the kidney: Implications for physiology and cadmium nephrotoxicity. *Metallomics* **2016**, *8*, 17–42. [CrossRef] [PubMed]

52. Thevenod, F.; Lee, W.K. Toxicology of cadmium and its damage to mammalian organs. In *Cadmium: From Toxicity to Essentiality*; Sigel, A., Sigel, H., Sigel, R.K.O., Eds.; Springer: Dordrecht, The Netherlands, 2013; Volume 11, pp. 415–490.

53. Wolff, N.A.; Ghio, A.J.; Garrick, L.M.; Garrick, M.D.; Zhao, L.; Fenton, R.A.; Thévenod, F. Evidence for mitochondrial localization of divalent metal transporter 1 (DMT1). *FASEB J.* **2014**, *28*, 2134–2145. [CrossRef] [PubMed]

54. Abouhamed, M.; Gburek, J.; Liu, W.; Torchalski, B.; Wilhelm, A.; Wolff, N.A.; Christensen, E.I.; Thevenod, F.; Smith, C.P. Divalent metal transporter 1 in the kidney proximal tubule is expressed in late endosomes/lysosomal membranes: Implications for renal handling of protein-metal complexes. *Am. J. Physiol. Ren. Physiol.* **2006**, *290*, F1525–F1533. [CrossRef] [PubMed]

55. Abouhamed, M.; Wolff, N.A.; Lee, W.K.; Smith, C.P.; Thevenod, F. Knockdown of endosomal/lysosomal divalent metal transporter 1 by RNA interference prevents cadmium-metallothionein-1 cytotoxicity in renal proximal tubule cells. *Am. J. Physiol. Ren. Physiol.* **2007**, *293*, F705–F712. [CrossRef] [PubMed]

56. Mitchell, C.J.; Shawki, A.; Ganz, T.; Nemeth, E.; Mackenzie, B. Functional properties of human ferroportin, a cellular iron exporter reactive also with cobalt and zinc. *Am. J. Physiol. Cell Physiol.* **2014**, *306*, C450–C459. [CrossRef] [PubMed]

57. Elinder, C.G.; Lind, B.; Kjellstorm, T.; Linnman, L.; Friberg, L. Cadmium in kidney cortex, liver and pancreas from Swedish autopsies: Estimation of biological half time in kidney cortex, considering calorie intake and smoking habits. *Arch. Environ. Health* **1976**, *31*, 292–301. [CrossRef] [PubMed]

58. Chung, J.; Nartey, N.O.; Cherian, M.G. Metallothionein levels in liver and kidney of Canadians—A potential indicator of environmental exposure to cadmium. *Arch. Environ. Health* **1986**, *41*, 319–323. [CrossRef] [PubMed]

59. Benedetti, J.L.; Samuel, O.; Dewailly, E.; Gingras, S.; Lefebvre, M.A. Levels of cadmium in kidney and liver tissues among a Canadian population (province of Quebec). *J. Toxicol. Environ. Health* **1999**, *56*, 145–163. [CrossRef] [PubMed]

60. Satarug, S.; Baker, J.R.; Reilly, P.E.B.; Moore, M.R.; Williams, D.J. Cadmium levels in the lung, liver, kidney cortex, and urine samples from Australians without occupational exposure to metals. *Arch. Environ. Health* **2002**, *57*, 69–77. [CrossRef] [PubMed]

61. Johansen, P.; Mulvad, G.; Pedersen, H.S.; Hansen, J.C.; Riget, F. Accumulation of cadmium in livers and kidneys in Greenlanders. *Sci. Total Environ.* **2006**, *372*, 58–63. [CrossRef] [PubMed]

62. Yoshida, M.; Ohta, H.; Yamauchi, Y.; Seki, Y.; Sagi, M.; Yamazaki, K.; Sumi, Y. Age-dependent changes in metallothionein levels in liver and kidney of the Japanese. *Biol. Trace Elem. Res.* **1998**, *63*, 167–175. [CrossRef] [PubMed]

63. Baba, H.; Tsuneyama, K.; Yazaki, M.; Nagata, K.; Minamisaka, T.; Tsuda, T.; Nomoto, K.; Hayashi, S.; Miwa, S.; Nakajima, T.; et al. The liver in itai-itai disease (chronic cadmium poisoning): Pathological features and metallothionein expression. *Mod. Pathol.* **2013**, *26*, 1228–1234. [CrossRef] [PubMed]

64. Uetani, M.; Kobayashi, E.; Suwazono, Y.; Honda, R.; Nishijo, M.; Nakagawa, H.; Kido, T.; Nogawa, K. Tissue cadmium (Cd) concentrations of people living in a Cd polluted area, Japan. *Biometals* **2006**, *19*, 521–525. [CrossRef] [PubMed]

65. Satarug, S.; Vesey, D.A.; Gobe, G.C. Health risk assessment of dietary cadmium intake: Do current guidelines indicate how much is safe? *Environ. Health Perspect.* **2017**, *125*, 284–288. [CrossRef] [PubMed]

66. Lyon, T.D.B.; Aughey, E.; Scott, R.; Fell, G.S. Cadmium concentrations in human kidney in the UK: 1978–1993. *J. Environ. Monit.* **1999**, *1*, 227–231. [CrossRef] [PubMed]

67. Dudley, R.E.; Gammal, L.M.; Klaassen, C.D. Cadmium-induced hepatic and renal injury in chronically exposed rats: Likely role of hepatic cadmium-metallothionein in nephrotoxicity. *Toxicol. Appl. Pharmacol.* **1985**, *77*, 414–426. [CrossRef]

68. Chan, H.M.; Zhu, L.-F.; Zhong, R.; Grant, R.; Goyer, R.A.; Cherian, M.G. Nephrotoxicity in rats following liver transplantation from cadmium-exposed rats. *Toxicol. Appl. Pharmacol.* **1993**, *123*, 89–96. [CrossRef] [PubMed]

69. Ellis, K.J.; Cohn, S.H.; Smith, T.J. Cadmium inhalation exposure estimates: Their significance with respect to kidney and liver burden. *J. Toxicol. Environ. Health* **1985**, *15*, 173–187. [CrossRef] [PubMed]

70. Barregard, L.; Fabricius-Lagging, E.; Lundh, T.; Mölne, J.; Wallin, M.; Olausson, M.; Modigh, C.; Sallsten, G. Cadmium, mercury, and lead in kidney cortex of living kidney donors: Impact of different exposure sources. *Environ. Res.* **2010**, *110*, 47–54. [CrossRef] [PubMed]

71. Akerstrom, M.; Barregard, L.; Lundh, T.; Sallsten, G. The relationship between cadmium in kidney and cadmium in urine and blood in an environmentally exposed population. *Toxicol. Appl. Pharmacol.* **2013**, *268*, 286–293. [CrossRef] [PubMed]

72. Satarug, S.; Swaddiwudhipong, W.; Ruangyuttikarn, W.; Nishijo, M.; Ruiz, P. Modeling cadmium exposures in low- and high-exposure areas in Thailand. *Environ. Health Perspect.* **2013**, *121*, 531–536. [CrossRef] [PubMed]

73. Crinnion, W.J. The CDC fourth national report on human exposure to environmental chemicals: What it tells us about our toxic burden and how it assists environmental medicine physicians. *Altern. Med. Rev.* **2010**, *15*, 101–108. [PubMed]

74. Riederer, A.M.; Belova, A.; George, B.J.; Anastas, P.T. Urinary cadmium in the 1999–2008 U.S. national health and nutrition examination survey (NHANES). *Environ. Sci. Technol.* **2013**, *47*, 1137–1147. [CrossRef] [PubMed]

75. Huang, M.; Choi, S.J.; Kim, D.W.; Kim, N.Y.; Bae, H.S.; Yu, S.D.; Kim, D.S.; Kim, H.; Choi, B.S.; Yu, I.J.; et al. Evaluation of factors associated with cadmium exposure and kidney function in the general population. *Environ. Toxicol.* **2013**, *28*, 563–570. [CrossRef] [PubMed]

76. Seo, J.W.; Kim, B.G.; Kim, Y.M.; Kim, R.B.; Chung, J.Y.; Lee, K.M.; Hong, Y.S. Trend of blood lead, mercury, and cadmium levels in Korean population: Data analysis of the Korea National Health and Nutrition Examination Survey. *Environ. Monit. Assess.* **2015**, *187*, 146. [CrossRef] [PubMed]

77. Garner, R.; Levallois, P. Cadmium levels and sources of exposure among Canadian adults. *Health Rep.* **2016**, *27*, 10–18. [PubMed]

78. Vacchi-Suzzi, C.; Kruse, D.; Harrington, J.; Levine, K.; Meliker, J.R. Is urinary cadmium a biomarker of long-term exposure in humans? A review. *Curr. Environ. Health Rep.* **2016**, *3*, 450–458. [CrossRef] [PubMed]

79. Kjellström, T.; Nordberg, G.F. A kinetic model of cadmium metabolism in the human being. *Environ. Res.* **1978**, *16*, 248–269. [CrossRef]

80. Elinder, C.G.; Kjellstorm, T.; Lind, B.; Molander, M.L.; Silander, T. Cadmium concentrations in human liver, blood, and bile: Comparison with a metabolic model. *Environ. Res.* **1978**, *17*, 236–241. [CrossRef]

81. Choudhury, H.; Harvey, T.; Thayer, W.C.; Lockwood, T.F. Urinary cadmium elimination as a biomarker of exposure for evaluating a cadmium dietary exposure-biokinetics model. *J. Toxicol. Environ. Health A* **2001**, *63*, 321–350. [CrossRef] [PubMed]

82. Diamond, G.L.; Thayer, W.C.; Choudhury, H. Pharmacokinetics/pharmacodynamics (PK/PD) modeling of risks of kidney toxicity from exposure to cadmium: Estimates of dietary risks in the U.S. population. *J. Toxicol. Environ. Health A* **2003**, *66*, 2141–2164. [CrossRef] [PubMed]

83. Ruiz, P.; Fowler, B.A.; Osterloh, J.D.; Fisher, J.; Mumtaz, M. Physiologically based pharmacokinetic (PBPK) tool kit for environmental pollutants—Metals. *SAR QSAR Environ. Res.* **2010**, *21*, 603–618. [CrossRef] [PubMed]

84. Ruiz, P.; Mumtaz, M.; Osterloh, J.; Fisher, J.; Fowler, B.A. Interpreting NHANES biomonitoring data, cadmium. *Toxicol. Lett.* **2010**, *198*, 44–48. [CrossRef] [PubMed]

85. Ruiz, P.; Ray, M.; Fisher, J.; Mumtaz, M. Development of a human physiologically based pharmacokinetic (PBPK) toolkit for environmental pollutants. *Int. J. Mol. Sci.* **2011**, *12*, 7469–7480. [CrossRef] [PubMed]

86. Amzal, B.; Julin, B.; Vahter, M.; Wolk, A.; Johanson, G.; Åkesson, A. Population toxicokinetic modeling of cadmium for health risk assessment. *Environ. Health Perspect.* **2009**, *117*, 1293–1301. [CrossRef] [PubMed]

87. Julin, B.; Vahter, M.; Amzal, B.; Wolk, A.; Berglund, M.; Åkesson, A. Relation between dietary cadmium intake and biomarkers of cadmium exposure in premenopausal women accounting for body iron stores. *Environ. Health* **2011**, *10*, 105. [CrossRef] [PubMed]

88. Béchaux, C.; Bodin, L.; Clémençon, S.; Crépet, A. PBPK and population modelling to interpret urine cadmium concentrations of the French population. *Toxicol. Appl. Pharmacol.* **2014**, *279*, 364–372. [CrossRef] [PubMed]

89. European Food Safety Authority (EFSA). Statement on tolerable weekly intake for cadmium. *EFSA J.* **2011**, *9*, 1975.

90. European Food Safety Authority (EFSA). Cadmium dietary exposure in the European population. *EFSA J.* **2012**, *10*, 2551.

91. Gobe, G.; Crane, D. Mitochondria, reactive oxygen species and cadmium toxicity in the kidney. *Toxicol. Lett.* **2010**, *198*, 49–55. [CrossRef] [PubMed]

92. Fujiwara, Y.; Lee, J.Y.; Tokumoto, M.; Satoh, M. Cadmium renal toxicity via apoptotic pathways. *Biol. Pharm. Bull.* **2012**, *35*, 1892–1897. [CrossRef] [PubMed]

93. Lenoir, O.; Tharaux, P.L.; Huber, T.B. Autophagy in kidney disease and aging: Lessons from rodent models. *Kidney Int.* **2016**, *90*, 950–964. [CrossRef] [PubMed]

94. Honda, R.; Swaddiwudhipong, W.; Nishijo, M.; Mahasakpan, P.; Teeyakasem, W.; Ruangyuttikarn, W.; Satarug, S.; Padungtod, C.; Nakagawa, H. Cadmium induced renal dysfunction among residents of rice farming area downstream from a zinc-mineralized belt in Thailand. *Toxicol. Lett.* **2010**, *198*, 26–32. [CrossRef] [PubMed]

95. Suwazono, Y.; Nogawa, K.; Morikawa, Y.; Nishijo, M.; Kobayashi, E.; Kido, T.; Nakagawa, H.; Nogawa, K. Renal tubular dysfunction increases mortality in the Japanese general population living in cadmium non-polluted areas. *J. Expo. Sci. Environ. Epidemiol.* **2015**, *25*, 399–404. [CrossRef] [PubMed]

96. Suwazono, Y.; Nogawa, K.; Morikawa, Y.; Nishijo, M.; Kobayashi, E.; Kido, T.; Nakagawa, H.; Nogawa, K. All-cause mortality increased by environmental cadmium exposure in the Japanese general population in cadmium non-polluted areas. *J. Appl. Toxicol.* **2015**, *35*, 817–823. [CrossRef] [PubMed]

97. Teeyakasem, W.; Nishijo, M.; Honda, R.; Satarug, S.; Swaddiwudhipong, W.; Ruangyuttikarn, W. Monitoring of cadmium toxicity in a Thai population with high-level environmental exposure. *Toxicol. Lett.* **2007**, *169*, 185–195. [CrossRef] [PubMed]

98. Gorriza, J.L.; Martinez-Castelao, A. Proteinuria: Detection and role in native renal disease progression. *Transplant. Rev.* **2012**, *26*, 3–13. [CrossRef] [PubMed]

99. Kobayashi, E.; Suwazono, Y.; Uetani, M.; Inaba, T.; Oishi, M.; Kido, T.; Nishijo, M.; Nakagawa, H.; Nogawa, K. Estimation of benchmark dose as the threshold levels of urinary cadmium, based on excretion of total protein, β2-microglobulin, and N-acetyl-β-D-glucosaminidase in cadmium non-polluted regions in Japan. *Environ. Res.* **2006**, *101*, 401–406. [CrossRef] [PubMed]

100. Kobayashi, E.; Suwazono, Y.; Uetani, M.; Kido, T.; Nishijo, M.; Nakagawa, H.; Nogawa, K. Tolerable level of lifetime cadmium intake estimated as a benchmark dose low, based on excretion of β2-microglobulin in the cadmium-polluted regions of the Kakehashi River Basin, Japan. *Bull Environ. Contam. Toxicol.* **2006**, *76*, 8–15. [CrossRef] [PubMed]

101. Kudo, K.; Konta, T.; Mashima, Y.; Ichikawa, K.; Takasaki, S.; Ikeda, A.; Hoshikawa, M.; Suzuki, K.; Shibata, Y.; Watanabe, T.; et al. The association between renal tubular damage and rapid renal deterioration in the Japanese population: The Takahata study. *Clin. Exp. Nephrol.* **2011**, *15*, 235–241. [CrossRef] [PubMed]

102. Mashima, Y.; Konta, T.; Kudo, K.; Takasaki, S.; Ichikawa, K.; Suzuki, K.; Shibata, Y.; Watanabe, T.; Kato, T.; Kawata, S.; et al. Increases in urinary albumin and β2-microglobulin are independently associated with blood pressure in the Japanese general population: The Takahata Study. *Hypertens. Res.* **2011**, *34*, 831–835. [CrossRef] [PubMed]

103. Pless-Mulloli, T.; Boettcher, M.; Steiner, M.; Berger, J. α-1-Microglobulin: Epidemiological indicator for tubular dysfunction induced by cadmium? *Occup. Environ. Med.* **1998**, *55*, 440–445. [CrossRef] [PubMed]

104. Ikeda, M.; Ezaki, T.; Tsukahara, T.; Moriguchi, J.; Furuki, K.; Fukui, Y.; Ukai, S.H.; Okamoto, S.; Sakurai, H. Critical evaluation of alpha1- and β2-microglobulins in urine as markers of cadmium-induced tubular dysfunction. *Biometals* **2004**, *17*, 539–541. [CrossRef] [PubMed]

105. Wallin, M.; Sallsten, G.; Lundh, T.; Barregard, L. Low-level cadmium exposure and effects on kidney function. *Occup. Environ. Med.* **2014**, *71*, 848–854. [CrossRef] [PubMed]

106. Ruangyuttikarn, W.; Panyamoon, A.; Nambunmee, K.; Honda, R.; Swaddiwudhipong, W.; Nishijo, M. Use of the kidney injury molecule-1 as a biomarker for early detection of renal tubular dysfunction in a population chronically exposed to cadmium in the environment. *SpringerPlus* **2013**, *2*, 533. [CrossRef] [PubMed]

107. Nishijo, M.; Satarug, S.; Honda, R.; Tsuritani, I.; Aoshima, K. The gender differences in health effects of environmental cadmium exposure and potential mechanisms. *Mol. Cell. Biochem.* **2004**, *255*, 87–92. [CrossRef] [PubMed]

108. Wu, X.; Jin, T.; Wang, Z.; Ye, T.; Kong, Q.; Nordberg, G. Urinary calcium as a biomarker of renal dysfunction in a general population exposed to cadmium. *J. Occup. Environ. Med.* **2001**, *43*, 898–904. [CrossRef] [PubMed]

109. Buchet, J.P.; Lauwerys, R.; Roels, H.; Bernard, A.; Bruaux, P.; Claeys, F.; Ducoffre, G.; de Plaen, P.; Staessen, J.; Amery, A.; et al. Renal effects of cadmium body burden of the general population. *Lancet* **1990**, *336*, 699–702. [CrossRef]

110. Thomas, L.D.; Hodgson, S.; Nieuwenhuijsen, M.; Jarup, L. Early kidney damage in a population exposed to cadmium and other heavy metals. *Environ. Health Perspect.* **2009**, *117*, 181–184. [CrossRef] [PubMed]

111. Prozialeck, W.C.; Van Dreel, A.; Ackerman, C.D.; Stock, I.; Papaeliou, A.; Yasmine, C.; Wilson, K.; Lamar, P.C.; Sears, V.L.; Gasiorowski, J.Z.; et al. Evaluation of cystatin C as an early biomarker of cadmium nephrotoxicity in the rat. *Biometals* **2016**, *29*, 131–146. [CrossRef] [PubMed]

112. Dieterle, F.; Perentes, E.; Cordier, A.; Roth, D.R.; Verdes, P.; Grenet, O.; Pantano, S.; Moulin, P.; Wahl, D.; Mahl, A.; et al. Urinary clusterin, cystatin C, β2-microglobulin and total protein as markers to detect drug-induced kidney injury. *Nat. Biotechnol.* **2010**, *28*, 463–469. [CrossRef] [PubMed]

113. Kuwata, K.; Nakamura, I.; Ide, M.; Sato, H.; Nishikawa, S.; Tanaka, M. Comparison of changes in urinary and blood levels of biomarkers associated with proximal tubular injury in rat models. *J. Toxicol. Pathol.* **2015**, *28*, 151–164. [CrossRef] [PubMed]

114. Argyropoulos, C.P.; Chen, S.S.; Ng, Y.H.; Roumelioti, M.E.; Shaffi, K.; Singh, P.P.; Tzamaloukas, A.H. Rediscovering β-2 microglobulin as a biomarker across the spectrum of kidney diseases. *Front. Med.* **2017**, *4*, 73. [CrossRef] [PubMed]

115. Liang, Y.; Lei, L.; Nilsson, J.; Li, H.; Nordberg, M.; Bernard, A.; Nordberg, G.F.; Bergdahl, I.A.; Jin, T. Renal function after reduction in cadmium exposure: An 8-year follow-up of residents in cadmium-polluted areas. *Environ. Health Perspect.* **2012**, *120*, 223–228. [CrossRef] [PubMed]

116. Kim, Y.D.; Yim, D.H.; Eom, S.Y.; Moon, S.I.; Park, C.H.; Kim, G.B.; Yu, S.D.; Choi, B.S.; Park, J.D.; Kim, H. Temporal changes in urinary levels of cadmium, N-acetyl-β-d-glucosaminidase and 2-microglobulin in individuals in a cadmium-contaminated area. *Environ. Toxicol. Pharmacol.* **2015**, *39*, 35–41. [CrossRef] [PubMed]

117. Crump, K.S. A new method for determining allowable daily intakes. *Fundam. Appl. Toxicol.* **1984**, *4*, 854–871. [CrossRef]

118. Gaylor, D.; Ryan, L.; Krewski, D.; Zhu, Y. Procedures for calculating benchmark doses for health risk assessment. *Regul. Toxicol. Pharmacol.* **1998**, *28*, 150–164. [CrossRef] [PubMed]

119. Ginsberg, G.L. Cadmium risk assessment in relation to background risk of chronic kidney disease. *J. Toxicol. Environ. Health* **2012**, *75*, 374–390. [CrossRef] [PubMed]

120. Suwazono, Y.; Sand, S.; Vahter, M.; Filipsson, A.F.; Skerfving, S.; Lidfeldt, J.; Akesson, A. Benchmark dose for cadmium-induced renal effects in humans. *Environ. Health Perspect.* **2006**, *114*, 1072–1076. [CrossRef] [PubMed]

121. Uno, T.; Kobayashi, E.; Suwazono, Y.; Okubo, Y.; Miura, K.; Sakata, K.; Okayama, A.; Ueshima, H.; Nakagawa, H.; Nogawa, K. Health effects of cadmium exposure in the general environment in Japan with special reference to the lower limit of the benchmark dose as the threshold level of urinary cadmium. *Scand. J. Work Environ. Health* **2005**, *31*, 307–315. [CrossRef] [PubMed]

122. Suwazono, Y.; Nogawa, K.; Uetani, M.; Nakada, S.; Kido, T.; Nakagawa, H. Application of the hybrid approach to the benchmark dose of urinary cadmium as the reference level for renal effects in cadmium polluted and non-polluted areas in Japan. *Environ. Res.* **2011**, *111*, 312–314. [CrossRef] [PubMed]

123. Hu, J.; Li, M.; Han, T.X.; Chen, J.W.; Ye, L.X.; Wang, Q.; Zhou, Y.K. Benchmark dose estimation for cadmium-induced renal tubular damage among environmental cadmium-exposed women aged 35–54 years in two counties of China. *PLoS ONE* **2014**, *9*, e115794. [CrossRef] [PubMed]

124. Ke, S.; Cheng, X.Y.; Zhang, J.Y.; Jia, W.J.; Li, H.; Luo, H.F.; Ge, P.H.; Liu, Z.M.; Wang, H.M.; He, J.S.; et al. Estimation of the benchmark dose of urinary cadmium as the reference level for renal dysfunction: A large sample study in five cadmium polluted areas in China. *BMC Public Health* **2015**, *15*, 656. [CrossRef] [PubMed]

125. Chen, L.; Jin, T.; Huang, B.; Nordberg, G.; Nordberg, M. Critical exposure level of cadmium for elevated urinary metallothionein-An occupational population study in China. *Toxicol. Appl. Pharmacol.* **2006**, *215*, 93–99. [CrossRef] [PubMed]

126. Chaumont, A.; Nickmilder, M.; Dumont, X.; Lundh, T.; Skerfving, S.; Bernard, A. Associations between proteins and heavy metals in urine at low environmental exposures: Evidence of reverse causality. *Toxicol. Lett.* **2012**, *210*, 345–352. [CrossRef] [PubMed]

127. Buser, M.C.; Ingber, S.Z.; Raines, N.; Fowler, D.A.; Scinicariello, F. Urinary and blood cadmium and lead and kidney function: NHANES 2007–2012. *Int. J. Hyg. Environ. Health* **2016**, *219*, 261–267. [CrossRef] [PubMed]

128. Hwangbo, Y.; Weaver, V.M.; Tellez-Plaza, M.; Guallar, E.; Lee, B.K.; Navas-Acien, A. Blood cadmium and estimated glomerular filtration rate in Korean adults. *Environ. Health Perspect.* **2011**, *119*, 1800–1805. [CrossRef] [PubMed]

129. De Nicola, L.; Zoccali, C. Chronic kidney disease prevalence in the general population: Heterogeneity and concerns. *Nephrol. Dial. Transplant.* **2016**, *31*, 331–335. [CrossRef] [PubMed]

130. Glassock, R.J.; David, G.; Warnock, D.G.; Delanaye, P. The global burden of chronic kidney disease: Estimates, variability and pitfalls. *Nat. Rev. Nephrol.* **2017**, *13*, 104–114. [CrossRef] [PubMed]

131. Crews, D.C.; Plantinga, L.C.; Miller, E.R.; Saran, R.; Hedgeman, E.; Saydah, S.H.; Williams, D.E.; Powe, N.R. Prevalence of chronic kidney disease in persons with undiagnosed or prehypertension in the United States. *Hypertension* **2010**, *55*, 1102–1109. [CrossRef] [PubMed]

132. Ferraro, P.M.; Costanzi, S.; Naticchia, A.; Sturniolo, A.; Gambaro, G. Low level exposure to cadmium increases the risk of chronic kidney disease: Analysis of the NHANES 1999–2006. *BMC Public Health* **2010**, *10*, 304. [CrossRef] [PubMed]

133. Navas-Acien, A.; Tellez-Plaza, M.; Guallar, E.; Muntner, P.; Silbergeld, E.; Jaar, B.; Weaver, V. Blood cadmium and lead and chronic kidney disease in US adults: A joint analysis. *Am. J. Epidemiol.* **2009**, *170*, 1156–1164. [CrossRef] [PubMed]

134. Lin, Y.S.; Ho, W.C.; Caffrey, J.L.; Sonawane, B. Low serum zinc is associated with elevated risk of cadmium nephrotoxicity. *Environ. Res.* **2014**, *134*, 133–138. [CrossRef] [PubMed]

135. Kim, N.H.; Hyun, Y.Y.; Lee, K.B.; Chang, Y.; Ryu, S.; Oh, K.H.; Ahn, C. Environmental heavy metal exposure and chronic kidney disease in the general population. *J. Korean Med. Sci.* **2015**, *30*, 272–277. [CrossRef] [PubMed]

136. Shi, Z.; Taylor, A.W.; Riley, M.; Byles, J.; Liu, J.; Noakes, M. Association between dietary patterns, cadmium intake and chronic kidney disease among adults. *Clin. Nutr.* **2017**, *5614*, 31366–31368. [CrossRef] [PubMed]

137. Scinicariello, F.; Abadin, H.G.; Murray, H.E. Association of low-level blood lead and blood pressure in NHANES 1999–2006. *Environ. Res.* **2011**, *111*, 1249–1257. [CrossRef] [PubMed]

138. Lee, B.K.; Kim, Y. Association of blood cadmium with hypertension in the Korean general population: analysis of the 2008–2010 Korean National Health and Nutrition Examination Survey data. *Am. J. Ind. Med.* **2012**, *55*, 1060–1067. [CrossRef] [PubMed]

139. Garner, R.E.; Levallois, P. Associations between cadmium levels in blood and urine, blood pressure and hypertension among Canadian adults. *Environ. Res.* **2017**, *155*, 64–72. [CrossRef] [PubMed]

140. Boonprasert, K.; Vesey, D.V.; Gobe, G.C.; Ruenweerayut, R.; Johnson, D.W.; Na-Bangchang, K.; Satarug, S. Is renal tubular cadmium toxicity clinically relevant? *Clin. Kidney J.* **2018**, 1–7. [CrossRef]

141. Padilla, M.A.; Elobeid, M.; Ruden, D.M.; Allison, D.B. An examination of the association of selected toxic metals with total and central obesity indices: NHANES 99-02. *Int. J. Environ. Res. Public Health* **2010**, *7*, 3332–3347. [CrossRef] [PubMed]

142. Jain, R.B. Effect of pregnancy on the levels of blood cadmium, lead, and mercury for females aged 17–39 years old: Data from National Health and Nutrition Examination Survey 2003–2010. *J. Toxicol. Environ. Health A* **2013**, *76*, 58–69. [CrossRef] [PubMed]

143. Shao, W.; Liu, Q.; He, X.; Liu, H.; Gu, A.; Jiang, Z. Association between level of urinary trace heavy metals and obesity among children aged 6–19 years: NHANES 1999–2011. *Environ. Sci. Pollut. Res. Int.* **2017**, *24*, 11573–11581. [CrossRef] [PubMed]

144. Dhooge, W.; Den Hond, E.; Koppen, G.; Bruckers, L.; Nelen, V.; Van De Mieroop, E.; Bilau, M.; Croes, K.; Baeyens, W.; Schoeters, G.; et al. Internal exposure to pollutants and body size in Flemish adolescents and adults: Associations and dose-response relationships. *Environ. Int.* **2010**, *36*, 330–337. [CrossRef] [PubMed]

145. Son, H.S.; Kim, S.G.; Suh, B.S.; Park, D.U.; Kim, D.S.; Yu, S.D.; Hong, Y.S.; Park, J.D.; Lee, B.K.; Moon, J.D.; et al. Association of cadmium with diabetes in middle-aged residents of abandoned metal mines: The first health effect surveillance for residents in abandoned metal mines. *Ann. Occup. Environ. Med.* **2015**, *27*, 13–20. [CrossRef] [PubMed]

146. Nie, X.; Wang, N.; Chen, Y.; Chen, C.; Han, B.; Zhu, C.; Chen, Y.; Xia, F.; Cang, Z.; Lu, M.; et al. Blood cadmium in Chinese adults and its relationships with diabetes and obesity. *Environ. Sci. Pollut. Res. Int.* **2016**, *23*, 18714–18723. [CrossRef] [PubMed]

147. Wilding, J.P.H. The role of the kidneys in glucose homeostasis in type 2 diabetes: Clinical implications and therapeutic significance through sodium glucose co-transporter 2 inhibitors. *Metab. Clin. Exp.* **2014**, *63*, 1228–1237. [CrossRef] [PubMed]

148. Vallon, V. The mechanisms and therapeutic potential of SGLT2 inhibitors in diabetes mellitus. *Annu. Rev. Med.* **2015**, *66*, 255–270. [CrossRef] [PubMed]

149. Reed, J.W. Impact of sodium–glucose cotransporter 2 inhibitors on blood pressure. *Vasc. Health Risk Manag.* **2016**, *12*, 393–405. [CrossRef] [PubMed]

150. Ellis, J.K.; Athersuch, T.J.; Thomas, L.D.; Teichert, F.; Pérez-Trujillo, M.; Svendsen, C.; Spurgeon, D.J.; Singh, R.; Järup, L.; Bundy, J.G.; et al. Metabolic profiling detects early effects of environmental and lifestyle exposure to cadmium in a human population. *BMC Med.* **2012**, *10*, 61. [CrossRef] [PubMed]

151. Suvagandha, D.; Nishijo, M.; Swaddiwudhipong, W.; Honda, R.; Ohse, M.; Kuhara, T.; Nakagawa, H.; Ruangyuttikarn, W. A biomarker found in cadmium exposed residents of Thailand by metabolome analysis. *Int. J. Environ. Res. Public Health* **2014**, *11*, 3661–3677. [CrossRef] [PubMed]

152. Thevenod, F.; Friedmann, J.M. Cadmium-mediated oxidative stress in kidney proximal tubule cells induces degradation of Na+/K+-ATPase through proteasomal and endo-/lysosomal proteolytic pathways. *FASEB J.* **1999**, *13*, 1751–1761. [CrossRef] [PubMed]

153. King, J.C.; Shames, D.M.; Woodhouse, L.R. Zinc homeostasis in humans. *J. Nutr.* **2000**, *130*, 1360S–1366S. [CrossRef] [PubMed]

154. Thijs, L.; Staessen, J.; Amery, A.; Bruaux, P.; Buchet, J.P.; Claeys, F.; De Plaen, P.; Ducoffre, G.; Lauwerys, R.; Lijnen, P. Determinants of serum zinc in a random population sample of four Belgian towns with different degrees of environmental exposure and body burden. *Environ. Health Perspect.* **1993**, *98*, 251–258. [CrossRef]

155. Pizent, A.; Jurasović, J.; Telisman, S. Serum calcium, zinc, and copper in relation to biomarkers of lead and cadmium in men. *J. Trace Elem. Med. Biol.* **2003**, *17*, 199–205. [CrossRef]

156. Satarug, S.; Nishijo, M.; Ujjin, P.; Moore, M.R. Chronic exposure to low-level cadmium induced zinc-copper dysregulation. *J. Trace Elem. Med. Biol.* **2018**, *46*, 32–38. [CrossRef] [PubMed]

157. Satarug, S.; Baker, J.R.; Reilly, P.E.; Moore, M.R.; Williams, D.J. Changes in zinc and copper homeostasis in human livers and kidneys associated with exposure to environmental cadmium. *Hum. Exp. Toxicol.* **2001**, *20*, 205–213. [CrossRef] [PubMed]

158. Boonprasert, K.; Satarug, S.; Morais, C.; Gobe, G.C.; Johnson, D.W.; Na-Bangchang, K.; Vesey, D.A. The stress response of human proximal tubule cells to cadmium involves up-regulation of haemoxygenase-1 and metallothionein but not cytochrome P450 enzymes. *Toxicol. Lett.* **2016**, *249*, 5–14. [CrossRef] [PubMed]

159. Garrett, S.H.; Sens, M.A.; Todd, J.H.; Somji, S.; Sens, D.A. Expression of MT-3 protein in the human kidney. *Toxicol. Lett.* **1999**, *105*, 207–214. [CrossRef]

160. Boonprasert, K.; Ruengweerayut, R.; Aunpad, R.; Satarug, S.; Na-Bangchang, K. Expression of metallothionein isoforms in peripheral blood leukocytes from Thai population residing in cadmium-contaminated areas. *Environ. Toxicol. Pharmacol.* **2012**, *34*, 935–940. [CrossRef] [PubMed]

161. Kayaalti, Z.; Mergen, G.; Söylemezoğlu, T. Effect of metallothionein core promoter region polymorphism on cadmium, zinc and copper levels in autopsy kidney tissues from a Turkish population. *Toxicol. Appl. Pharmacol.* **2010**, *245*, 252–255. [CrossRef] [PubMed]

162. Kayaalti, Z.; Aliyev, V.; Söylemezoğlu, T. The potential effect of metallothionein 2A-5A/G single nucleotide polymorphism on blood cadmium, lead, zinc and copper levels. *Toxicol. Appl. Pharmacol.* **2011**, *256*, 1–7. [CrossRef] [PubMed]

163. Hattori, Y.; Naito, M.; Satoh, M.; Nakatochi, M.; Naito, H.; Kato, M.; Takagi, S.; Matsunaga, T.; Seiki, T.; Sasakabe, T.; et al. Metallothionein MT2A A-5G polymorphism as a risk factor for chronic kidney disease and diabetes: Cross-sectional and cohort studies. *Toxicol. Sci.* **2016**, *152*, 181–193. [CrossRef] [PubMed]

Article

Urinary Cadmium Threshold to Prevent Kidney Disease Development

Soisungwan Satarug [1,2], Werawan Ruangyuttikarn [3], Muneko Nishijo [4] and Patricia Ruiz [5,*]

[1]   National Research Centre for Environmental Toxicology, the University of Queensland,
     Brisbane 4108, Australia; sj.satarug@yahoo.com.au
[2]   UQ Diamantina Institute and Centre for Health Services Research, Centre for Kidney Disease Research and
     Translational Research Institute, Woolloongabba, Brisbane 4102, Australia
[3]   Division of Toxicology, Department of Forensic Medicine, Chiang Mai University,
     Chiang Mai 50200, Thailand; ruangyuttikarn@gmail.com
[4]   Department of Public Health, Kanazawa Medical University, Uchinada, Ishikawa 920-0293, Japan;
     ni-koei@kanazawa-med.ac.jp
[5]   Computational Toxicology and Methods Development Laboratory, Division of Toxicology and Human
     Health Sciences, Agency for Toxic Substances and Disease Registry, Centers for Disease Control and
     Prevention, Atlanta, GA 30333, USA
*   Correspondence: pruiz@cdc.gov

Received: 9 March 2018; Accepted: 23 April 2018; Published: 1 May 2018

Abstract: The frequently observed association between kidney toxicity and long-term cadmium (Cd) exposure has long been dismissed and deemed not to be of clinical relevance. However, Cd exposure has now been associated with increased risk of developing chronic kidney disease (CKD). We investigated the link that may exist between kidney Cd toxicity markers and clinical kidney function measure such as estimated glomerular filtration rates (eGFR). We analyzed data from 193 men to 202 women, aged 16−87 years [mean age 48.8 years], who lived in a low- and high-Cd exposure areas in Thailand. The mean (range) urinary Cd level was 5.93 (0.05–57) μg/g creatinine. The mean (range) for estimated GFR was 86.9 (19.6−137.8) mL/min/1.73 m². Kidney pathology reflected by urinary β2-microglobulin (β2-MG) levels $\geq$ 300 μg/g creatinine showed an association with 5.32-fold increase in prevalence odds of CKD ($p = 0.001$), while urinary Cd levels showed an association with a 2.98-fold greater odds of CKD prevalence ($p = 0.037$). In non-smoking women, Cd in the highest urinary Cd quartile was associated with 18.3 mL/min/1.73 m² lower eGFR value, compared to the lowest quartile ($p < 0.001$). Evidence for Cd-induced kidney pathology could thus be linked to GFR reduction, and CKD development in Cd-exposed people. These findings may help prioritize efforts to reassess Cd exposure and its impact on population health, given the rising prevalence of CKD globally.

Keywords: β2-microglobulin; cadmium; chronic kidney disease; clinical kidney function measure; estimated glomerular filtration rate; *N*-acetyl-β-D-glucosaminidase; population health; tubular dysfunction; toxicity threshold limit; urine protein

1. Introduction

Exposure to the heavy metal cadmium (Cd) is inevitable for most people as this metal is present in foodstuffs, cigarette smoke and polluted air [1–4]. By total diet studies, staple foods such as rice, potatoes, and wheat constitute 40–60% of total dietary Cd intake in the average consumer in various populations [4]. In addition, offal, spinach, shellfish, crustacean and mollusks constitute dietary Cd sources [4]. Cd oxide (CdO) in cigarette smoke and polluted air has relatively high bioavailability. Consequently, most smokers show elevated Cd levels in their blood, urine, and tissues [1–4]. To-date, non-occupational Cd exposure has been associated with numerous chronic diseases of continuously rising prevalence, notably type-2 diabetes [2–4]. However, the most frequently reported Cd toxicity in non-occupationally exposed populations is related to kidneys, notably the injury to the proximal tubular epithelial cells that reabsorb and concentrate Cd from the glomerular filtrate [1–4]. Renal tubular cells are highly susceptible to Cd-induced apoptosis because of high abundance of mitochondria and substantial reliance on autophagy to maintain homeostasis [5–7]. One of the consequential results of the injury and death of renal tubular epithelial cells by Cd is a reduction in tubular reabsorption capacity in Cd-exposed people, leading to loss of nutrients through urine, notably glucose, amino acids, calcium, and zinc [8–11].

Urinary levels of *N*-acetyl-β-D-glucosaminidase (NAG) enzyme and the low molecular weight protein β2-microglobulin (β2-MG) are often used to reflect Cd-induced kidney tubular pathologies [12–15], while urinary Cd excretion is used as an indicator of cumulative long-term exposure or body burden [16–18]. For example, elevated urinary β2-MG levels ($\geq$283 μg/day) were reported for subjects who excreted 3.05 μg of Cd per day [8]. In Japanese studies, urinary Cd levels 1.6–4.6 μg/g creatinine were associated with urinary β2-MG levels $\geq$ 1000 μg/g creatinine, an indicative of severe and irreversible tubular dysfunction [19,20]. However, these signs of Cd-related tubular toxicity have not been considered to be clinically relevant.

Challenging the notion on a lack of clinical and health risk implications are data from the representative of Korean population, and from two cycles of the U.S. National Health and Nutrition Examination Surveys (NHANES) showing that Cd exposure may increase the risk of developing chronic kidney disease (CKD) [21–24]. Dietary Cd intake has also been associated with CKD development in Chinese population [25]. However, none of these studies has assessed glomerular filtration rate (GFR) concurrently with kidney tubular pathology markers, notably urine NAG and β2-MG. We hypothesize that GFR falls as the result of Cd destroys tubular cells after reabsorption from filtrate. Hence, the present study investigated the potential link between Cd tubular toxicity and CKD in Cd-exposed Thai subjects. We sought to evaluate an independent association between GFR reduction and evidence of tubular pathologies in relation to urinary Cd levels, age, gender, body mass index (BMI), and smoking.

2. Methods

2.1. Study Subjects

To represent chronic environmental exposure situations, we assembled a group of 395 subjects from a low-Cd exposure area in Bangkok [26], and high-Cd exposure area in rural rice farming villages in Mae Sot District, known to be an area with Cd contamination [12,27]. Subjects in neither low-Cd nor high-Cd exposure group were occupationally exposed to metals. The Institutional Ethical Committee, Chulalongkorn University Hospital, approved the Bangkok study protocol, while the Mae Sot Hospital Ethical Committee approved the Mae Sot study protocol. All participants provided informed consent prior to participation. For a low-Cd exposure group, inclusion criteria were apparently healthy. Exclusion criteria were pregnancy, breast-feeding, history of metal work, a hospital record or diagnosis by physician of CKD, heart disease, diabetes, anemia, or hyperlipidemia. For a high-Cd exposure group, subjects were randomly selected from 13 villages with cadmium pollution in Mae Sot District, Tak Province. There were cases of diagnosed CKD, hypertension, and osteoporosis as shown in Table 1. Smoking, regular use of medications, education, occupation, family health history and anthropometric data were obtained from questionnaires. Excluding those with incomplete data, 395 subjects (180 from the Bangkok group, and 215 from the Mae Sot group) form study subjects in the present study.

Table 1. Characteristic of study subjects.

Descriptors/Variables	All Subjects $n = 395$	Men $n = 193$	Women $n = 202$	*p* Values
Age (years)	48.8 ± 14.0	47.4 ± 15.8	50.1 ± 12.0	0.024
BMI (kg/m^2)	22.2 ± 3.8	22.0 ± 3.4	22.4 ± 4.1	0.387
Smoking prevalence (%) [a]	45.1	66.8	24.3	<0.001
Hypertension (%) [b]	21.7	24.2	19.3	0.240
eGFR (mL/min/1.73 m^2)	86.9 ± 24.2	87.2 ± 25.0	86.6 ± 23.5	0.717
CKD prevalence (%)	12.7	13	12.4	0.863
Exposure indicators				
Urinary creatinine (mg/dL)	100.2 ± 67.7	115.1 ± 71.7	85.8 ± 60.5	<0.001
Urinary Cd concentration (µg/L)	6.65 ± 10.70	7.48 ± 12.71	5.87 ± 8.29	0.930
Urinary Cd (µg/g creatinine)	5.93 ± 7.69	5.43 ± 7.60	6.41 ± 7.77	0.061
Urinary Cd >1 µg/g creatinine (%) [c]	55.9	53.4	58.4	0.312
Urinary Cd >5.24 µg/g creatinine (%) [d]	40.3	37.3	43.1	0.243
Renal pathology markers				
β2-MG (mg/g creatinine)	2.68 ± 12.43	3.35 ± 13.87	2.04 ± 10.88	0.973
β2-MG ≥ 1 mg/g creatinine (%) [e]	14.2	17.1	11.4	0.104
NAG (Units/ g creatinine)	5.31 ± 4.26	4.98 ± 3.50	5.63 ± 4.86	0.103
Total protein (mg/g creatinine)	75.6 ± 142	74.7 ± 144	76.4 ± 141	0.200
Reported health status (%)				
No disease	66.7	67.4	66.0	0.901
Anemia	6.2	4.7	7.6	0.221
Hypertension	15.5	16.3	14.7	0.796
Diabetes	2.8	3.2	2.5	0.763
Osteoporosis	3.1	0.5	5.6	0.004
Kidney disease	3.1	4.7	1.5	0.083
Urinary stones	1.6	1.6	1.5	1.000
Others	1.0	1.6	0.5	0.317

Numbers are arithmetic mean ± standard deviation (SD). eGFR is determined with CKD−EPI equation, and eGFR < 60 mL/min/1.73 m^2 is defined as CKD [28]. [a] Both current and ex-smokers are grouped together because of a known long half-life of Cd in the body. [b] Hypertension was defined as systolic blood pressure ≥ 140 mmHg, or diastolic blood pressure ≥ 90 mmHg, physician diagnosis, or prescription of anti-hypertensive medications. [c] Tubular Cd toxicity threshold, established by the European Food Safety Agency [29]. [d] Tubular Cd toxicity threshold, established by the FAO/WHO [30]. [e] Severe and irreversible tubular dysfunction [19,20].

2.2. Ascertainment of Long-Term Cadmium Exposure Levels

Assessment of long-term Cd exposure or body burden was based on creatinine-adjusted urinary Cd concentrations. Urinary Cd is a suitable exposure marker to assess kidney effects since the majority of Cd in urine is ultrafilterable, but not reabsorbed by kidney tubules [2]. The plasma Cd concentration reflects Cd influx into blood circulation from external sources (diet and air) and internal reservoirs (liver). Accordingly, urinary Cd excretion rate is proportional to plasma Cd concentrations, glomerular filtration rates and tubular sequestration rates [2]. For the Bangkok group [26], the urinary Cd concentrations were determined with the inductively-coupled plasma/mass spectrometry, calibrated with multi-element standards (EM Science, EM Industries Inc., Newark, NJ, USA). Quality assurance and control were conducted with simultaneous analysis of samples of the reference urine Lyphochek® (Bio-Rad, Sydney, Australia), which contained low- and high-range Cd levels. The coefficient of variation of 2.5% was obtained for Cd in the reference urine. Cd concentrations of urine samples reported below the limit of detection (LOD) of 0.05 µg/L were assigned as the LOD divided by the square root of 2. For the Mae Sot group [13], urinary Cd concentrations were determined with an atomic absorption spectrophotometer (Shimadzu Model AA-6300, Kyoto, Japan). Urine standard reference material No. 2670 (The National Institute of Standards, Washington, DC, USA) was used for quality assurance and control purposes.

2.3. Clinical Kidney Function Measure and Assessment of Tubular and Glomerular Integrity

Clinical kidney function measure was based on estimated glomerular filtration rate (eGFR), calculated with the Chronic Kidney Disease Epidemiology Collaboration (CKD-EPI) equation [28]. Male eGFR = 141 × [serum creatinine ÷ 0.9]^Y × 0.993^{age}, where Y = −0.411 if serum creatinine ≤ 0.9 mg/dL, Y= −1.209 if serum creatinine > 0.9 mg/dL. Female eGFR = 144 × [serum creatinine÷0.7]^Y × 0.993^{age}, where Y= −0.329 if serum creatinine ≤ 0.7 mg/dL, Y= −1.209 if serum creatinine > 0.7 mg/dL. CKD is defined as eGFR < 60 mL/min/1.73 m^2, and CKD stages I, II, III, IV and V correspond to eGFR 90–119, 60–89, 30–59, 15–29 and <15 mL/min/1.73 m^2, respectively [28].

Assessment of tubular dysfunction was based on a reduction in tubular reabsorption activity, reflected by an increase in urinary excretion rate of β2-MG [12–15]. Due to a small molecular weight, β2-MG is filtered, reabsorbed by tubules, and approximately 0.3% of filtered β2-MG is excreted in urine [2]. Assessment of tubular integrity was based on urinary excretion of the enzyme NAG [12–15] and urinary NAG excretion is considered to be proportional to nephron numbers as this enzyme originates mostly from tubular epithelial cells which is released upon cell injury [2]. For the Bangkok group, the urinary β2-MG assay was based on the latex immunoagglutination method (LX test, Eiken 2MGII; Eiken and Shionogi Co., Tokyo, Japan), and the urinary NAG assay was based on an enzymatic reaction and colorimetry. The urinary protein assay was based on turbidimetry (Roche/Hitachi 717, Boehringer Mannheim and Roche Diagnostics, Roche Diagnostics GmbH Mannheim, Germany). The urinary and serum creatinine assay was based on the Jaffe's reaction.

For Mae Sot group, the urinary β2-MG assay was based on an enzyme immunoassay (GLAZYME β2 microglobulin-EIA test kit, Sanyo Chemical Industries, Ltd., Kyoto, Japan), while the urinary NAG assayed was based on colorimetry (NAG test kit, Shionogi Pharmaceuticals, Sapporo, Japan). The urinary protein assay was based on the Kingsbury-Clark method, while the urinary and serum creatinine assay was based on the Jaffe's reaction.

2.4. Statistical Analysis

The SPSS statistical package 17.0 (SPSS Inc., Chicago, IL, USA) was used to analyze data. We used the Mann-Whitney U-test to compare two groups of subjects. The distribution of the variables was examined for skewness and those showing right skewing were subjected to logarithmic transformation before analysis, where required. One sample Kolmogorov-Smirnov test was used to detect a departure from normal distribution of variables. We used the logistic regression analysis to estimate Prevalence Odds Ratio (POR) for CKD, attributable to Cd exposure and kidney tubular pathologies. The univariate analysis was used to estimate effect size of Cd exposure levels with adjustment for covariates and urinary Cd quartiles × smoking × gender interactions. In addition, we used a multilinear regression analysis to evaluate the strength of associations between eGFR and its predictors in subjects stratified by gender, smoking status and Cd exposure levels. *p* values ≤ 0.05 for a two-tailed test was considered to indicate statistical significance.

3. Results

3.1. Characteristics of Study Subjects

Of 395 study subjects, 202 were women and 193 were men. The mean age of women was 4 years older than the mean age of men of 47.4 years (*p* = 0.024). Smoking was more prevalent in men than women (66.8% vs. 24.3%) (*p* < 0.001). The mean (SD) values for eGFR were 86.9 (24.2) mL/min/1.78 m^2 (range: 19.6–137.8). The CKD prevalence was 13% in men and 12.4% in women (*p* = 0.863), while hypertension prevalence was 24.2% in men and 19.7% in women (*p* = 0.240).

The mean urinary creatinine concentrations in men was higher than women (*p* < 0.001). The mean urinary Cd concentrations in men (7.48 µg/L) and women (5.87 µg/L) did not differ (*p* = 0.930). The mean (SD) urinary Cd was 5.93 (7.69) µg/g creatinine (range: 0.05–57.57). The mean urinary Cd tended to be higher in women than men, when data were adjusted for urine dilution by creatinine excretion (6.41 vs. 5.43 µg/g creatinine, $p = 0.061$). The prevalence of urinary Cd levels above 5.24 µg/g creatinine was 40.3%, while more than half (55.9%) of the subjects had urinary Cd levels, exceeding 1 µg/g creatinine. The prevalence of severe and irreversible tubular dysfunction (urinary β2-MG levels $\geq$ 1000 µg/g creatinine) was 17.1% in men and 14. 2% in women ($p = 0.104$). Urinary β2-MG, NAG and protein levels in men and women did not differ.

In all subjects, creatinine-adjusted urinary Cd levels showed a strong correlation with age (Spearman rank's correction coefficient (r) = 0.644, p = < 0.001), and this association between age and urinary Cd levels persisted after stratification by smoking status ($r = 0.627$, $p < 0.001$ for non-smokers, $r = 0.540$, $p < 0.001$ for smokers). There was an inverse correlation between urinary Cd levels and BMI ($r = -0.214$, $p < 0.005$) in all subjects. After controlling for age, the association of urinary Cd levels and BMI persisted in smokers only ($r = -0.166$, $p = 0.027$), while there was a tendency for an association in non-smokers ($r = -0.119$, $p = 0.081$).

The prevalence rates of various diseases reported by participants differed in men and women (Likelihood Chi-square 15.5, $p = 0.03$). Osteoporosis was more prevalent in women than men (5.6% vs. 0.5%, $p = 0.004$). Kidney disease diagnosis tended to be higher in men than women (4.7% vs. 1.5%, $p = 0.083$).

3.2. CKD Prevalence Associated with Tubular Dysfunction and Cadmium Exposure

By logistic regression analysis of CKD prevalence (Table 2), an increase in CKD prevalence odds was found to be associated with age ($p < 0.001$), BMI ($p = 0.001$), tubular dysfunction (urinary β2-MG levels $\geq$ 300 µg/g creatinine) ($p = 0.001$), urinary Cd ($p = 0.037$) and protein levels ($p = 0.023$). Elevated β2-MG levels associated with the highest increase in CKD prevalence odds (POR 5.324, 95% CI: 2.035, 13.928), followed by urinary Cd levels (POR 2.978, 95% CI: 1.066, 8.317), urinary protein (POR 1.900, 95% CI: 1.093, 3.302), BMI (POR 1.188, 95% CI: 1.071, 1.318), and age (POR 1.119, 95%CI: 1.070, 1.170). Urine NAG did not associate with CKD prevalence ($p = 0.744$).

Table 2. Increased prevalence odds of chronic kidney disease associated with cadmium and severity of tubular dysfunction.

Independent Variables	POR of CKD	95% CI for POR		p Values
		Lower	Upper	
Gender	0.771	0.317	1.876	0.566
Age (years)	1.119	1.070	1.170	<0.001
BMI (kg/m^2)	1.188	1.071	1.318	0.001
Smoking	1.002	0.378	2.661	0.996
Tubular dysfunction [a]	5.324	2.035	13.928	0.001
Log urine Cd (µg/g creatinine)	2.978	1.066	8.317	0.037
Log urine NAG (units/g creatinine)	1.340	0.231	7.770	0.744
Log urine protein (mg/g creatinine)	1.900	1.093	3.302	0.023

POR = Prevalence Odds Ratio. [a] Tubular dysfunction is defined as urinary β2-MG levels $\geq$ 300 µg/g creatinine [19,20]. POR was derived from a logistic regression model analysis in which CKD (eGFR < 60 mL/min/1.73 m^2) was a categorical dependent variable, while age, BMI, creatinine adjusted urinary Cd, NAG and protein levels were continuous independent variables. Categorical independent variables were gender, smoking status, and tubular dysfunction. p values $\leq$ 0.05 are considered to indicate statistical significant levels.

3.3. Effect Size Estimates

Table 3 provides results of a univariate analysis that quantified the variation in eGFR attributable to various independent variables and their interactions. Factors and covariates in the first column accounted for more than a half (67. 3%, $p < 0.001$) of the total eGFR variation. Age accounted for the largest proportion (36.4%) of eGFR variability ($p < 0.001$), while BMI, Cd quartiles, and urine β2-MG each accounted for 2.7% ($p = 0.001$), 5.1% ($p < 0.001$) and 3.3% ($p < 0.001$) in eGFR variation among study subjects. Gender, smoking, urine NAG and protein did not contribute significantly to eGFR variation. There was a significant interaction between gender × smoking that contributed to 2.9% (p = 0.011) of eGFR variation. There was a tendency for Cd quartiles × smoking interaction (0.8%, p = 0.076).

Table 3. Univariate analysis of glomerular filtration rates.

Factors and Covariates	Degree of Freedom	eGFR (mL/min/1.73 m²)		
		F	p	η^2
Corrected Model	19	43.715	<0.001	0.689
Intercept	1	219.823	<0.001	0.370
Age (years)	1	214.578	<0.001	0.364
BMI (kg/m²)	1	10.484	0.001	0.027
Smoking	1	0.082	0.775	0.000
Gender	1	1.347	0.247	0.004
Log urine β2-MG (µg/g creatinine)	1	12.800	<0.001	0.033
Log urine NAG (units/g creatinine)	1	0.275	0.600	0.001
Log urine protein (mg/g creatinine)	1	2.405	0.122	0.006
Urinary Cd quartiles	3	6.765	<0.001	0.051
Gender × smoking	1	3.161	0.076	0.008
Urinary Cd quartiles × smoking	3	3.747	0.011	0.029
Urinary Cd quartiles × gender	3	0.890	0.446	0.007
Urinary Cd quartiles × gender × smoking	2	2.032	0.133	0.011
Error	375			
Total	395			
Corrected Total	394			

Adjusted R^2 = 0.673; η^2 = eta squared. Adjusted R^2 value describes the total eGFR variability attributable to all factors and covariates. The η^2 value describes the proportion of eGFR variability attributable to each factor/covariate. eGFR in mL/min/1.73 m² was a continuous dependent variable. Age, BMI, creatinine-adjusted urinary β2-MG, NAG, and protein excretion levels were continuous independent variables. Gender, smoking status and Cd exposure levels were categorical independent variables. p values $\leq$ 0.05 are considered to indicate statistical significant levels. Quartiles 1, 2, 3, and 4 of Cd exposure levels correspond to urinary Cd; 0.05–0.50, 0.51–2.95, 2.96–8.80, 8.81–57.57 µg/g creatinine, respectively. The numbers of subjects in quartile 1, 2, 3 and 4 were 100, 101, 97, and 97, respectively.

Adjusted mean eGFR across urinary Cd quartiles 1, 2, 3 and 4 were shown separately for male and female non-smokers (Figure 1), given a significant gender × smoking interaction (Table 3). eGFR reduction was associated with urinary Cd levels in a dose-dependent manner in non-smoking women, but not in men. The adjusted mean eGFR [SE] values for urinary Cd quartiles 1, 2, 3 and 4 in non-smoking women were 97.6 [2.3], 95.8 [2.1], 82.9 [2.7], and 79.3 [2.5] mL/min/1.73 m², respectively. The adjusted mean eGFR [SE] in urinary quartile 4, 3 and 2 was 18.3 [3.5] (p < 0.001), 14.6 [3.6] (p = 0.005) and 1.7 [2.9] (p = 1.000) mL/min/1.73 m² lower than the adjusted mean eGFR in urinary Cd quartile 1, respectively. The total number of non-smoking women was 153 and the numbers (%) distribution in urinary Cd quartiles 1, 2, 3 and 4 were 42 (27.5%), 49 (32%), 27 (17.6%), 35 (22.9%), respectively. The total number of non-smoking men was 64 and the number (%) distribution in urinary Cd quartile 1, 2, 3 and 4 were 36 (56.3%), 16 (25%), 7 (10.9%), and 5 (7.8%), respectively. In non-smoker male group, adjusted mean eGFR, the adjusted mean eGFR [SE] in urinary quartile 4, 3 and 2 was 20.7 [6.8] (p = 0.193), 21.1 [7.1] (p = 0.251) and 19.5 [7.6] (p = 1.000) mL/min/1.73 m² lower than the adjusted mean eGFR in urinary Cd quartile 1, respectively.

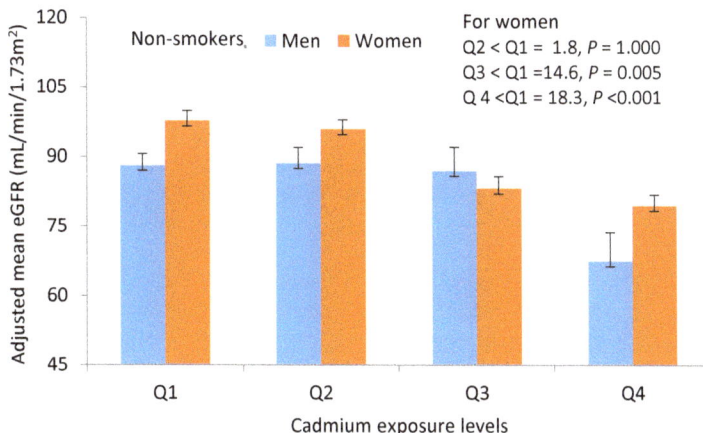

Figure 1. Cadmium-dose dependent reduction in glomerular filtration rates in non-smoking women. Bars represent mean eGFR $\pm$ SE values for groups of subjects stratified according to the quartiles of urinary Cd excretion levels. Urinary Cd levels in quartiles 1, 2, 3 and 4 are 0.05–0.50, 0.51–2.95, 2.96–8.80, 8.81–57.57 µg/g creatinine, and the corresponding numbers of subjects are 100, 101, 97 and 97, respectively. The mean eGFR values are adjusted for interactions and covariates as follows; age 48.79 years, BMI 22.21 kg/m^2, urinary β2-MG 59.74 µg/g creatinine, NAG 4.29 units/g creatinine, and protein excretion 24.98 mg/g creatinine. *p* values $\leq$ 0.05 indicate statistically significant difference between adjusted mean eGFR in quartile 2, 3 and 4, compared with the urinary Cd quartile 1.

3.4. Evidence for Urinary Cd Threshold Level

Table 4 shows results of a multilinear regression, used to further explore associations of eGFR, and kidney pathology markers. Age, BMI, gender, smoking, urinary Cd, β2-MG, NAG, and protein levels accounted for 66.5% of the total eGFR variation among study subjects. Age showed the strongest inverse association with eGFR (β = −0.548, *p* < 0.001), followed by urine Cd (β = −0.234, *p* < 0.001), β2-MG (β = −0.178, *p* < 0.001) and BMI (β = −0.105, *p* = 0.001). There was a marginal association between eGFR and female gender (β = 0.066, *p* = 0.051). Associations of eGFR and urinary NAG (β = 0.004, *p* = 0.893) and protein levels (β = −0.037, *p* = 0.236) were not significant.

Table 4. Multilinear regression analysis of glomerular filtration rates.

Independent Variables	eGFR (mL/min/1.73 m^2)			
	Standardized β coefficients	95% CI for β		*p* Value
		Lower	Upper	
Age (years)	−0.548	−1.084	−0.812	<0.001
BMI (kg/m^2)	−0.105	−1.068	−0.288	0.001
Gender	0.066	−0.017	6.435	0.051
Smoking	−0.002	−3.543	3.397	0.967
Log urine Cd (µg/g creatinine)	−0.234	−10.900	−4.976	<0.001
Log urine β2-MG (µg/g creatinine)	−0.178	−4.773	−1.808	<0.001
Log urine NAG (units/g creatinine)	0.004	−5.042	5.780	0.893
Log urine protein (mg/g creatinine)	−0.037	−2.982	0.736	0.236

Adjusted R^2 = 0.665, *p* < 0.001. eGFR was a continuous dependent variable. Gender (male = 1, female = 2), smoking (non-smoker = 1, smoker = 2) were categorical independent variables, while age, BMI, creatinine adjusted urinary Cd, NAG and protein levels were continuous independent variables. *p* values $\leq$ 0.05 are considered to indicate statistical significant levels.

Figures 2–4 provide data for the strength (β) of associations of eGFR and kidney pathology markers (urine β2-MG, NAG, and protein) across urinary Cd quartiles. In all subjects (Figure 2), a strongly inverse association was seen between eGFR and β2-MG in both non-smokers (β = −0.486, $p < 0.001$) and smokers (β = −0.619, $p < 0.001$). In unadjusted models, eGFR was not associated with β2-MG in Cd exposure quartile 1 (β = 0.058, $p = 0.570$), but in quartile 2 (β = −0.295, $p = 0.003$), quartile 3 (β = −0.545, $p < 0.001$) and quartile 4 (β = −0.650, $p < 0.001$). After adjustment for covariates and interactions; the β coefficients (p values) of the eGFR and β2-MG association in Cd exposure quartiles 1, 2, 3 and 4 were 0.013 ($p = 0.897$), −0.246 ($p = 0.020$), −0.547 ($p < 0.001$), and −0.685 ($p < 0.001$), respectively.

In all subjects (Figure 3), a marginally inverse association between eGFR and NAG was seen in non-smokers (β = −0.189, $p = 0.005$), while a moderately inverse association was in smokers (β = −0.396, $p < 0.001$). In unadjusted models, a marginally positive association between eGFR and NAG was evident in urinary Cd quartile 1 (β = 0.206, $p = 0.039$), eGFR and NAG association was absent in quartile 2 (β = −0.020, $p = 0.845$). A strongly inverse association was seen between eGFR and NAG in quartile 3 (β = −0.471, $p < 0.001$), while a moderately inverse association existed in quartile 4 (β = −0.265, $p = 0.009$). After adjustment for covariates and interactions, the β coefficient (p value) of eGFR and NAG association in Cd exposure quartiles 1, 2, 3 and 4 were 0.190 ($p = 0.057$), −0.061 ($p = 0.546$), −0.445 ($p < 0.001$), and −0.271, ($p = 0.009$), respectively.

In all subjects (Figure 4), a marginally positive association was seen between eGFR and urinary protein in non-smokers (β = 0.144, $p = 0.034$), but a moderately inverse association was seen in smokers (β = −0.280, $p < 0.001$). In unadjusted models, an association of eGFR and urine protein was not present in urinary Cd quartile 1 ($p = 0.629$) and quartile 2 ($p = 0.912$), while a moderately inverse association was evident in quartile 3 (β = −0.359, $p < 0.001$) and quartile 4 (β = −0.399, $p < 0.001$), After adjustment for covariates and interactions, the β coefficients (p values) of eGFR and protein associations in urinary Cd quartiles 1, 2, 3 and 4 were 0.062 ($p = 0.565$), 0.013 ($p = 0.894$), −0.350 ($p < 0.001$), and −0.413 ($p < 0.001$), respectively.

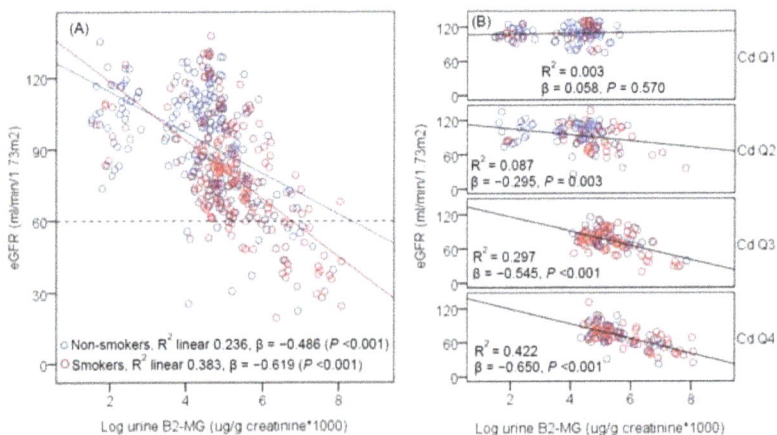

Figure 2. Scatterplots of kidney function measure vs. tubular dysfunction biomarker. The regression lines of eGFR vs. urinary β2-MG levels are shown for groups of subjects according to smoking status (**A**) and urinary Cd quartiles (**B**). The reference line in (**A**) is based on the CKD diagnosis, eGFR < 60 mL/min/1.73 m². The R^2 values and the β coefficients shown in (**A**,**B**) are unadjusted. Urinary Cd levels in quartiles 1, 2, 3 and 4 are 0.05–0.50, 0.51–2.95, 2.96–8.80, and 8.81–57.57 µg/g creatinine, and the corresponding numbers of subjects are 100, 101, 97 and 97, respectively.

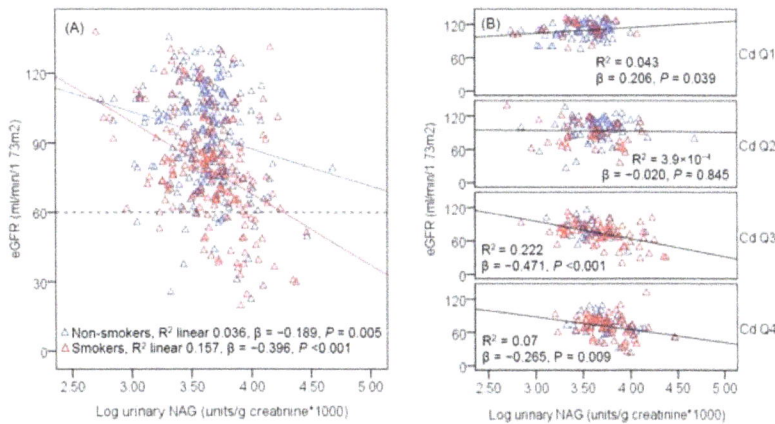

Figure 3. Scatterplots of kidney function measure vs. tubular injury biomarker. The regression lines of eGFR vs. urinary NAG levels are shown for groups of subjects according to smoking status (**A**) and urinary Cd quartiles (**B**). The reference line in (**A**) is based on the CKD diagnosis, eGFR < 60 mL/min/1.73 m². The R^2 values the β coefficients shown in (**A**,**B**) are unadjusted. Urinary Cd levels in quartiles 1, 2, 3 and 4 are 0.05–0.50, 0.51–2.95, 2.96–8.80, and 8.81–57.57 µg/g creatinine, and the corresponding numbers of subjects are 100, 101, 97 and 97, respectively.

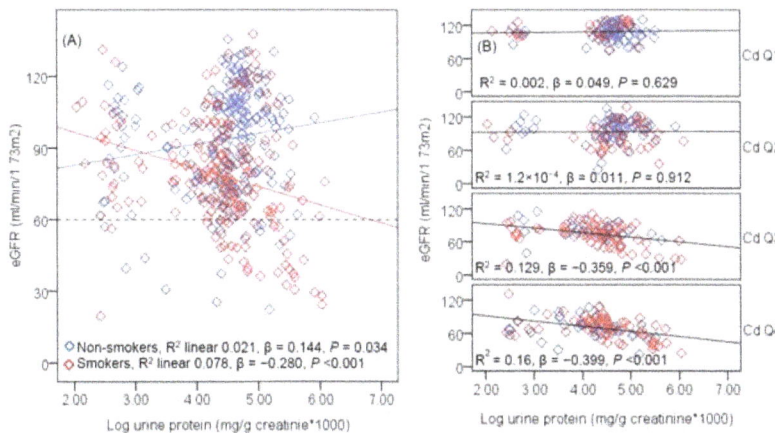

Figure 4. Scatterplots of kidney function measure vs. glomerular damage biomarker. The regression lines of eGFR vs. urinary protein levels are shown for subjects stratified according to smoking status (**A**) and urinary Cd quartiles (**B**). The reference line in (**A**) is based on the CKD diagnosis, eGFR < 60 mL/min/1.73 m². The R^2 values and the β coefficients shown in (**A**,**B**) are unadjusted. Urinary Cd levels in quartiles 1, 2, 3 and 4 are 0.05–0.50, 0.51–2.95, 2.96–8.80, and 8.81–57.57 µg/g creatinine, and the corresponding numbers of subjects are 100, 101, 97 and 97, respectively.

4. Discussion

Herein, we have observed for the first time an association of a 5.32-fold rise in CKD prevalence odds and urinary β2-MG levels ≥ 300 µg/g creatinine in Thai subjects with chronic environmental exposure to Cd. This independent association between elevated levels of urinary β2-MG and a marked increase in odds of CKD prevalence suggests a vital role played by kidney tubular cells in the pathogenesis and/or progression of CKD. Indeed, a tubular-glomerular connection is increasingly recognized [31] as is the evidence for β2-MG as marker of a range of kidney disease [32–34]. Our finding concurs with experimental data and clinical outcomes that suggest urinary β2-MG is a predictor of GFR reduction [32–34].

Supporting tubular-glomerular connection are data from a prospective cohort study in Japan showing that a sign of tubular impairment (urine β2-MG levels ≥ 300 μg/g creatinine) was associated with a 79% (95% CI: 1.07, 2.99) increase in the likelihood of having eGFR fall at high rates, i.e., 10 mL/min/1.73 m^2 over 5-year observation period [35]. In another cross-sectional study, a milder tubular impairment (urine β2-MG levels ≥ 145 μg/g creatinine) was associated with an increase in the prevalence odds for hypertension in Japanese subjects [36]. Results of these Japanese studies underscored clinical values of urine β2-MG measurement, but Cd exposure levels experienced by Japanese subjects in these two studies were not measured. Thus, it is unknown if these observed outcomes (rapid GFR reduction and hypertension development) in subjects with high urine β2-MG levels could be linked to Cd or other environmental factors.

Urinary Cd levels > 1 μg/L (>0.5 μg/g creatinine) were associated with a 48% increase in the risk of CKD development (95% CI: 1.01, 2.17) in adult participants in the U.S. NHANES 1999–2006 cycle [21]. Consistent with the U.S. study is our finding of an association between elevated Cd body burden, assessed by urinary Cd levels, and an increase in odds of CKD prevalence (2.98 fold). Multilinear regression data indicated also that lower eGFR values were associated with higher urinary Cd levels. Further, in an effect-size analysis, a dose-response between eGFR reduction and urinary Cd quartiles was evident in non-smoking women. This may implicate dietary Cd intake in the pathogenesis of CKD. Likewise, in a Chinese population study, cumulative Cd intake estimate was associated with a 4-fold increase in CKD prevalence (95% CI: 2.91, 5.63) [25].

An association of lower eGFR and higher blood Cd levels was noted in Korean population [37] and the representative of the U.S. population (the U.S. NHANES, 2007–2012) [38]. In a Korean study, blood Cd levels in the highest tertile were associated with 1.85 mL/min/1.73 m^2 lower GFR values (95% CI: −3.55, −0.16), compared with the lowest tertile [37]. However, it is noteworthy the majority of Cd in blood is in red blood cells, which are not filtered (not present in glomerular filtrate) [2]. Consequently, it is impossible to attribute blood Cd to eGFR reduction and to Cd toxicity in the kidney in the absence of data on kidney pathology. A 2.91-fold increase in CKD risk (95% CI: 1.76, 4.81) was associated with blood Cd levels > 0.6 μg/L in the U.S. NHANES 1999–2006 adult participants [23]. Blood Cd levels > 0.53 μg/L were associated with an approximately two-fold increase in risk of CKD development (95% CI: 1.09, 4.50) among adult participants in the NHANES 2011–2012 [22]. In the Korean population study, elevated blood Cd levels, but not blood Pb or blood Hg, were associated with CKD, especially in those with hypertension [24].

Currently, urinary Cd threshold limit for CKD is lacking. However, there are several urinary Cd threshold limits that have been derived for kidney tubular toxicity using benchmark dose method [2,39]. In one study, urinary Cd 0.57–1.84 μg/g creatinine was identified as threshold levels for urinary β2-MG levels ≥ 1065 μg/g creatinine [14]. As discussed above, it was evident that GFR and CKD both were substantially associated with elevated urine β2-MG and that association of GFR and β2-MG was minimal (or absent) in subjects with urinary Cd in quartile 1 (urinary Cd 0.05–0.50 μg/g creatinine). In the absence of threshold limit for CKD and a continuously rising CKD prevalence worldwide, it is argued that urinary Cd of 0.50 μg/g creatinine might be useful. Urinary Cd of 0.50 μg/g creatinine is 2-fold and 10-fold lower than the threshold level for kidney Cd toxicity, established by the European Food Safety Agency [29] and the WHO/FAO [30], respectively. Urinary Cd levels < 1 μg/g creatinine have been found to be associated with kidney pathologies in many previous studies [3,4]. In a study of Swedish women, 53–64 years of age, urinary Cd of 0.67 μg/g and 0.8 μg/g creatinine were found to be associated with markers of tubular impairment and glomerular dysfunction, respectively [40]. Urinary Cd of 0.74 μg/g creatinine was associated with albuminuria in the Torres Strait (Australia) women who had diabetes [41].

5. Conclusions

For the first time, we have demonstrated that a clinical kidney function measure such as estimated glomerular filtration rates could be linked to both Cd exposure and tubular toxicity in Cd-dose and toxicity severity dependent manner. In addition, we have shown that a urinary Cd level as low as 0.50 µg/g creatinine might be used as a warning sign of excessive Cd intake, Cd toxic burden, kidney pathologies and kidney function deterioration. Urinary Cd of 0.50 µg/g creatinine is 10-fold lower than current threshold for kidney toxicity established by the FAO/WHO of 5.24 µg/g creatinine. This established urinary Cd threshold level does not afford health protection. Consequently, there is an urgent need to reassess Cd toxic burden and urinary Cd toxicity threshold limit that should prevent human population from excessive Cd exposure, and CKD development.

6. Strengths and Limitations

The strengths of this study include the samples of men and women with homogeneous exposure sources (i.e., none were occupationally exposed) together with a wide Cd-exposure range (urinary Cd 0.05–58 µg/g creatinine) and a wide eGFR range (19.6–137.8 mL/min/1.73 m^2) suitable for dose-response relationship analysis. The high CKD prevalence of 12.7% in villages with varying degrees contamination allowed recruitment of sufficient numbers of subjects with low GFR and CKD. The community-based recruitment strategy minimized bias toward certain subpopulation groups, frequently encountered in health center-based studies.

The limitations of this study were its small sample size and its cross-sectional design, which limited an assessment of temporal relationships between variables or causal inference of Cd exposure. A wide age range was another limitation as GFR falls with increasing age due to loss of nephrons [2]. GFR could also fall due to tubular pathologies induced by Cd and other environmental nephrotoxicants. Most subjects with high-Cd exposure were rice farmers, co-exposure to other nephrotoxicants in pesticides might also be a possible confounder. Heavy smoking, and presence of disease notably hypertension and diabetes were likely confounders. GFR may fall because of kidney damage due to smoking. This was evident in Figure 2, where an additional effect of smoking on eGFR was suggested by the increasing β slope in urinary Cd quartiles 3 and 4, relative to quartiles 1 and 2, given the higher prevalence of smokers in urinary quartile 3 (35.4%) and quartile 4 (32%), compared to quartile 2 (20.2%) and quartile 1 (12.4%).

GFR may also fall because of kidney damage due to hypertension, and because of nephron loss, urinary excretion of NAG in heavy smokers, hypertensive and diabetic subjects could be lower than expected. This was evident in Figure 3, where there was a marked drop in the β slope in quartile 4, compared with quartile 3. Such a drop in β slope could be interpreted to be resulted from loss of tubular cells, leading to lower urinary NAG excretion levels than expected in quartile 4.

Urinary Cd concentrations were determined by two methods. For low-Cd exposure group, a high sensitive and high specificity method, known as inductively-coupled plasma mass spectrometry, was used. A less sensitive, but sufficiently high specificity assay with atomic absorption spectrophotometer was used for high-Cd exposure group. However, data from quality control and assurance conduced with standard urine specimens suggest that variation due to different methods was relatively small.

Author Contributions: S.S., W.R. and M.N. designed study protocols. S.S. and W.R. obtained ethical institutional clearances for research on human subjects and supervised biologic specimen collection in Thailand. S.S., W.R. and M.N. supervised biologic specimen analysis in Australia and Japan. S.S., W.R., M.N., and P.R. analyzed and interpreted data. S.S., P.R. wrote and revised the manuscript.

Acknowledgments: S.S. was a recipient of the Reverse Brain Drain Awards (2003–2006) from the Commission for High Education, Thailand Ministry of Education. The National Research Centre for Environmental Toxicology was funded by Queensland Health, the University of Queensland, Queensland University of Technology, and Griffith University.

Conflicts of Interest: The authors declare no conflict of interest.

Disclaimer: The findings and conclusions in this presentation have not been formally disseminated by [the Centers for Disease Control and Prevention/the Agency for Toxic Substances and Disease Registry] and should not be construed to represent any agency determination or policy.

References

1. Agency for Toxic Substances and Disease Registry (ATSDR). Toxicological Profile for Cadmium, Department of Health and Humans Services, Public Health Service, Centers for Disease Control and Prevention, Atlanta, GA, USA. 2012. Available online: http://www.atsdr.cdc.gov/toxprofiles/tp5.pdf (accessed on 8 March 2018).
2. Satarug, S. Dietary cadmium intake and its effects on kidneys. *Toxics* **2018**, *6*. [CrossRef] [PubMed]
3. Satarug, S.; Vesey, D.A.; Gobe, G.C. Current health risk assessment practice for dietary cadmium: Data from different countries. *Food Chem. Toxicol.* **2017**, *106*, 430–445. [CrossRef] [PubMed]
4. Satarug, S.; Vesey, D.A.; Gobe, G.C. Health risk assessment of dietary cadmium intake: Do current guidelines indicate how much is safe? *Environ. Health Perspect.* **2017**, *125*, 284–288. [CrossRef] [PubMed]
5. Lenoir, O.; Tharaux, P.L.; Huber, T.B. Autophagy in kidney disease and aging: Lessons from rodent models. *Kidney Int.* **2016**, *90*, 950–964. [CrossRef] [PubMed]
6. Gobe, G.; Crane, D. Mitochondria, reactive oxygen species and cadmium toxicity in the kidney. *Toxicol. Lett.* **2010**, *198*, 49–55. [CrossRef] [PubMed]
7. Fujiwara, Y.; Lee, J.Y.; Tokumoto, M.; Satoh, M. Cadmium renal toxicity via apoptotic pathways. *Biol. Pharm. Bull.* **2012**, *35*, 1892–1897. [CrossRef] [PubMed]
8. Buchet, J.P.; Lauwerys, R.; Roels, H.; Bernard, A.; Bruaux, P.; Claeys, F.; Ducoffre, G.; de Plaen, P.; Staessen, J.; Amery, A.; et al. Renal effects of cadmium body burden of the general population. *Lancet* **1990**, *336*, 699–702. [CrossRef]
9. Wu, X.; Jin, T.; Wang, Z.; Ye, T.; Kong, Q.; Nordberg, G. Urinary calcium as a biomarker of renal dysfunction in a general population exposed to cadmium. *J. Occup. Environ. Med.* **2001**, *43*, 898–904. [CrossRef] [PubMed]
10. Nishijo, M.; Satarug, S.; Honda, R.; Tsuritani, I.; Aoshima, K. The gender differences in health effects of environmental cadmium exposure and potential mechanisms. *Mol. Cell Biochem.* **2004**, *255*, 87–92. [CrossRef] [PubMed]
11. Satarug, S.; Nishijo, M.; Ujjin, P.; Moore, M.R. Chronic exposure to low-level cadmium induced zinc-copper dysregulation. *J. Trace Elem. Med. Biol.* **2018**, *46*, 32–38. [CrossRef] [PubMed]
12. Teeyakasem, W.; Nishijo, M.; Honda, R.; Satarug, S.; Swaddiwudhipong, W.; Ruangyuttikarn, W. Monitoring of cadmium toxicity in a Thai population with high-level environmental exposure. *Toxicol. Lett.* **2007**, *169*, 185–195. [CrossRef] [PubMed]
13. Honda, R.; Swaddiwudhipong, W.; Nishijo, M.; Mahasakpan, P.; Teeyakasem, W.; Ruangyuttikarn, W.; Satarug, S.; Padungtod, C.; Nakagawa, H. Cadmium induced renal dysfunction among residents of rice farming area downstream from a zinc-mineralized belt in Thailand. *Toxicol. Lett.* **2010**, *198*, 26–32. [CrossRef] [PubMed]
14. Hu, J.; Li, M.; Han, T.X.; Chen, J.W.; Ye, L.X.; Wang, Q.; Zhou, Y.K. Benchmark dose estimation for cadmium-induced renal tubular damage among environmental cadmium-exposed women aged 35–54 years in two counties of China. *PLoS ONE* **2014**, *9*, e115794. [CrossRef] [PubMed]
15. Wallin, M.; Sallsten, G.; Lundh, T.; Barregard, L. Low-level cadmium exposure and effects on kidney function. *Occup. Environ. Med.* **2014**, *71*, 848–854. [CrossRef] [PubMed]
16. Satarug, S.; Baker, J.R.; Reilly, P.E.; Moore, M.R.; Williams, D.J. Cadmium levels in the lung, liver, kidney cortex, and urine samples from Australians without occupational exposure to metals. *Arch. Environ. Health* **2002**, *57*, 69–77. [CrossRef] [PubMed]
17. Akerstrom, M.; Barregard, L.; Lundh, T.; Sallsten, G. The relationship between cadmium in kidney and cadmium in urine and blood in an environmentally exposed population. *Toxicol. Appl. Pharmacol.* **2013**, *268*, 286–293. [CrossRef] [PubMed]
18. Ikeda, M.; Nakatsuka, H.; Watanabe, T.; Shimbo, S. Estimation of daily cadmium intake from cadmium in blood or cadmium in urine. *Environ. Health Prev. Med.* **2015**, *20*, 455–459. [CrossRef] [PubMed]

19. Kobayashi, E.; Suwazono, Y.; Uetani, M.; Kido, T.; Nishijo, M.; Nakagawa, H.; Nogawa, K. Tolerable level of lifetime cadmium intake estimated as a benchmark dose low, based on excretion of β2-microglobulin in the cadmium-polluted regions of the Kakehashi River Basin, Japan. *Bull. Environ. Contam. Toxicol.* **2006**, *76*, 8–15. [CrossRef] [PubMed]

20. Liang, Y.; Lei, L.; Nilsson, J.; Li, H.; Nordberg, M.; Bernard, A.; Nordberg, G.F.; Bergdahl, I.A.; Jin, T. Renal function after reduction in cadmium exposure: An 8-year follow-up of residents in cadmium-polluted areas. *Environ. Health Perspect.* **2012**, *120*, 223–228. [CrossRef] [PubMed]

21. Ferraro, P.M.; Costanzi, S.; Naticchia, A.; Sturniolo, A.; Gambaro, G. Low level exposure to cadmium increases the risk of chronic kidney disease: Analysis of the NHANES 1999–2006. *BMC Public Health* **2010**, *10*, 304. [CrossRef] [PubMed]

22. Lin, Y.S.; Ho, W.C.; Caffrey, J.L.; Sonawane, B. Low serum zinc is associated with elevated risk of cadmium nephrotoxicity. *Environ. Res.* **2014**, *134*, 133–138. [CrossRef] [PubMed]

23. Navas-Acien, A.; Tellez-Plaza, M.; Guallar, E.; Muntner, P.; Silbergeld, E.; Jaar, B.; Weaver, V. Blood cadmium and lead and chronic kidney disease in US adults: A joint analysis. *Am. J. Epidemiol.* **2009**, *170*, 1156–1164. [CrossRef] [PubMed]

24. Kim, N.H.; Hyun, Y.Y.; Lee, K.B.; Chang, Y.; Ryu, S.; Oh, K.H.; Ahn, C. Environmental heavy metal exposure and chronic kidney disease in the general population. *J. Korean Med. Sci.* **2015**, *30*, 272–277. [CrossRef] [PubMed]

25. Shi, Z.; Taylor, A.W.; Riley, M.; Byles, J.; Liu, J.; Noakes, M. Association between dietary patterns, cadmium intake and chronic kidney disease among adults. *Clin. Nutr.* **2017**, *5614*, 31366–31368. [CrossRef] [PubMed]

26. Satarug, S.; Nishijo, M.; Ujjin, P.; Vanavanitkun, Y.; Moore, M.R. Cadmium-induced nephropathy in the development of high blood pressure. *Toxicol. Lett.* **2005**, *157*, 57–68. [CrossRef] [PubMed]

27. Satarug, S.; Swaddiwudhipong, W.; Ruangyuttikarn, W.; Nishijo, M.; Ruiz, P. Modeling cadmium exposures in low- and high-exposure areas in Thailand. *Environ. Health Perspect.* **2013**, *12*, 531–536. [CrossRef] [PubMed]

28. Levey, A.S.; Coresh, J.; Bolton, K.; Culleton, B.; Harvey, K.S.; Ikizler, T.A.; Johnson, C.A.; Kausz, A.; Kimmel, P.L.; Kusek, J.; et al. K/DOQI clinical practice guidelines for chronic kidney disease: Evaluation, classification and stratification. *Am. J. Kidney Dis.* **2002**, *39*, S1–S266.

29. European Food Safety Authority (EFSA). Statement on tolerable weekly intake for cadmium. *EFSA J.* **2011**, *9*. Available online: http://www.efsa.europa.eu/en/efsajournal/doc/1975.pdf (accessed on 8 March 2018).

30. FAO/WHO. Food and Agriculture Organization of the United Nations. In Proceedings of the Seventy-third Meeting, Geneva, Switzerland, 8–17 June 2010; Summary and Conclusions. Available online: http://www.who.int/foodsafety/publications/chem/summary73.pdf (accessed on 8 March 2018).

31. Nakhoul, N.; Batuman, V. Role of proximal tubules in the pathogenesis of kidney disease. *Contrib. Nephrol.* **2011**, *169*, 37–50. [PubMed]

32. Dieterle, F.; Perentes, E.; Cordier, A.; Roth, D.R.; Verdes, P.; Grenet, O.; Pantano, S.; Moulin, P.; Wahl, D.; Mahl, A.; et al. Urinary clusterin, cystatin C, β2-microglobulin and total protein as markers to detect drug-induced kidney injury. *Nat. Biotechnol.* **2010**, *28*, 463–469. [CrossRef] [PubMed]

33. Kuwata, K.; Nakamura, I.; Ide, M.; Sato, H.; Nishikawa, S.; Tanaka, M. Comparison of changes in urinary and blood levels of biomarkers associated with proximal tubular injury in rat models. *J. Toxicol. Pathol.* **2015**, *28*, 151–164. [CrossRef] [PubMed]

34. Argyropoulos, C.P.; Chen, S.S.; Ng, Y.H.; Roumelioti, M.E.; Shaffi, K.; Singh, P.P.; Tzamaloukas, A.H. Rediscovering beta-2 microglobulin as a biomarker across the spectrum of kidney diseases. *Front. Med.* **2017**, *4*, 73. [CrossRef] [PubMed]

35. Kudo, K.; Konta, T.; Mashima, Y.; Ichikawa, K.; Takasaki, S.; Ikeda, A.; Hoshikawa, M.; Suzuki, K.; Shibata, Y.; Watanabe, T.; et al. The association between renal tubular damage and rapid renal deterioration in the Japanese population: The Takahata study. *Clin. Exp. Nephrol.* **2011**, *15*, 235–241. [CrossRef] [PubMed]

36. Mashima, Y.; Konta, T.; Kudo, K.; Takasaki, S.; Ichikawa, K.; Suzuki, K.; Shibata, Y.; Watanabe, T.; Kato, T.; Kawata, S.; et al. Increases in urinary albumin and beta2-microglobulin are independently associated with blood pressure in the Japanese general population: The Takahata Study. *Hypertens. Res.* **2011**, *34*, 831–835. [CrossRef] [PubMed]

37. Hwangbo, Y.; Weaver, V.M.; Tellez-Plaza, M.; Guallar, E.; Lee, B.K.; Navas-Acien, A. Blood cadmium and estimated glomerular filtration rate in Korean adults. *Environ. Health Perspect.* **2011**, *119*, 1800–1805. [CrossRef] [PubMed]
38. Buser, M.C.; Ingber, S.Z.; Raines, N.; Fowler, D.A.; Scinicariello, F. Urinary and blood cadmium and lead and kidney function: NHANES 2007–2012. *Int. J. Hyg. Environ. Health* **2016**, *219*, 261–267. [CrossRef] [PubMed]
39. Ginsberg, G.L. Cadmium risk assessment in relation to background risk of chronic kidney disease. *J. Toxicol. Environ. Health* **2012**, *75*, 374–390. [CrossRef] [PubMed]
40. Akesson, A.; Lundh, T.; Vahter, M.; Bjellerup, P.; Lidfeldt, J.; Nerbrand, C.; Samsioe, G.; Strömberg, U.; Skerfving, S. Tubular and glomerular kidney effects in Swedish women with low environmental cadmium exposure. *Environ. Health Perspect.* **2005**, *113*, 1627–1631. [CrossRef] [PubMed]
41. Haswell-Elkins, M.; Imray, P.; Satarug, S.; Moore, M.R.; O'dea, K. Urinary excretion of cadmium among Torres Strait Islanders (Australia) at risk of elevated dietary exposure through traditional foods. *J. Expo. Sci. Environ. Epidemiol.* **2007**, *17*, 372–377. [CrossRef] [PubMed]

toxics

MDPI

Article

Cancer Mortality in Residents of the Cadmium-Polluted Jinzu River Basin in Toyama, Japan

Muneko Nishijo [1,2,*], Hideaki Nakagawa [3], Yasushi Suwazono [4], Kazuhiro Nogawa [4],
Masaru Sakurai [5], Masao Ishizaki [5] and Teruhiko Kido [6]

1   Department of Public Health, Kanazawa Medical University, Uchinada, Ishikawa 920-0293, Japan
2   Health Evaluation Center, Kanazawa Medical University Hospital, Uchinada, Ishikawa 920-0293, Japan
3   Medical Research Institute, Kanazawa Medical University, Uchinada, Ishikawa 920-0293, Japan;
    hnakagaw@kanazawa-med.ac.jp
4   Department of Occupational and Environmental Medicine, Graduate School of Medicine, Chiba University,
    Chiba 260-8670, Japan; suwa@faculty.chiba-u.jp (Y.S.); nogagwa@chiba-u.jp (K.N.)
5   Department of Hygiene, Kanazawa Medical University, Uchinada, Ishikawa 920-0293, Japan;
    m-sakura@kanazawa-med.ac.jp (M.S.); issa1@kanazawa-med.ac.jp (M.I.)
6   Department of Community Health Nursing, School of Health Sciences, Kanazawa University,
    Kanazawa 920-0942, Japan; kido@mns.mp.kanazawa-u.ac.jp
*   Correspondence: ni-koei@kanazawa-med.ac.jp; Tel.: +81-76-218-8430

Received: 28 February 2018; Accepted: 4 April 2018; Published: 6 April 2018

Abstract: After 26 years, we followed up 7348 participants in a 1979–1984 health screening survey in the Jinzu River basin, the heaviest cadmium-polluted area in Japan. We assessed the associations of cadmium exposure levels and mortality from cancer and renal damage, indicated by records of proteinuria and glucosuria in the original survey. Mortality risks (hazard ratios) were analyzed using the Cox proportional hazards model, stratified by sex, after adjusting for age, smoking status, and hypertension, as indicated in the original survey records. In men, the adjusted hazard ratio for mortality from lung cancer was significantly lower in individuals residing in an area of historically high cadmium exposure and in subjects with a historical record of proteinuria, glucosuria, and glucoproteinuria. The risk of mortality from prostate cancer was borderline higher in cadmium-exposed men. In women, historical cadmium exposure was not associated with an increased risk of mortality from malignant neoplasms, but the adjusted hazard ratios for death from total malignant neoplasms or from renal and uterine cancers were significantly higher in exposed subjects with a historical record of proteinuria, glucosuria, and glucoproteinuria. These findings suggest that women residing in cadmium-polluted areas who exhibit markers of renal damage may be at risk of dying of cancer.

Keywords: cadmium; follow-up study; cause of death; mortality; environmental pollution; cancer

1. Introduction

Cadmium (Cd) compounds have been classified as human carcinogens by the International Agency for Research on Cancer [1], leading to studies of mortality causes in Cd-exposed populations. Nawrot et al. (2006) reported a significant association between Cd exposure and lung cancer risk in a Belgian cohort, suggesting that aspiration of house dust containing contaminated soil particles may be related to an increase in the incidence of lung cancer [2]. In male factory workers exposed to high levels of Cd, increased incidences of prostate cancer were observed in Sweden [3] and the UK [4]. An increased breast cancer risk was reported in US women with higher urinary Cd levels [5] and in Swedish women whose dietary Cd intake was high [6]. Akesson et al. (2008) reported that dietary Cd intake increased postmenopausal endometrial cancer incidence in a Swedish cohort [7]. In the general American population, increased mortality from lung and pancreatic cancers in men, and from ovarian and uterine cancers in women, were suggested to be associated with urinary Cd [8]. Also, a significant association between Cd exposure and renal cancer has been reported [9].

In our previous study in residents of the Cd-exposed Jinzu River basin in Toyama, Japan [10], the mortality risks for cancers of the colon and rectum, uterus, and kidney and urinary tract were significantly higher in exposed women with glucoproteinuria. These findings suggested increased cancer risks associated with renal damage induced by Cd exposure. The small sample, however, precluded an analysis of mortality risks for specific cancers.

In this study, we investigated the associations between Cd exposure and mortality risks for specific cancers, and analyzed mortality risks in Jinzu River basin residents with historical records of renal damage (indicated by proteinuria, glucosuria, and glucoroteinuria).

2. Materials and Methods

2.1. Study Subjects

The Jinzu River basin, the largest and most Cd-polluted area in Japan, is an endemic area of itai-itai disease, which is prevalent among older women and characterized by osteomalacia with severe bone pain and renal tubular dysfunction [11,12]. Cd exposure in this area was divided into five levels (no exposure and borderline, mild, moderate, and high exposure) based on the contribution of contaminated Jinzu River water to irrigation water in each area. The prevalence of itai-itai disease or glucoproteinuria in women aged over 50 years has previously been shown to be associated with Cd exposure [13]. After the discovery of itai-itai disease, the Japanese government conducted health screenings in six Cd-polluted areas, including the Jinzu River basin, to identify residents with renal damage.

A total of 7348 participants (3363 men and 3985 women) in the 1979–1984 health screening survey in Toyama, who constituted 97.6% of the 7531 residents aged over 50, were targeted in the present follow-up survey. These participants lived in the Cd-polluted Jinzu River basin areas of Toyama City, Fuchu, Ohsawano, and Yatsuo. Non-polluted sections of two towns and five cities were selected as controls, and a total of 2098 residents (926 men and 1172 women) participated in the health screenings. These controls were also used in the present survey. Table 1 shows the age distribution, smoking status, and hypertension among the exposed and control subjects, obtained by questionnaire during the original survey [14]. The participants in the exposed areas were divided into two groups: the renal dysfunction group with proteinuria, glucosuria, and glucoproteinuria (807 men and 801 women), and the healthy resident group with neither glucosuria nor proteinuria (2556 men and 3184 women).

After the first step of the baseline health screening test, urinary Cd and urinary beta2-microglobulin (β2-MG) were measured in the subjects with proteinuria, glucosuria, and glucoproteinuria. Table 2 shows medians with the 90–95th of urinary Cd and positive rate (%) of urinary β2-MG in groups with different urinary findings: only glucosuria, only proteinuria, glucosuria and proteinuria. The 95th percentiles of urinary Cd of all groups were more than 20 (μg/L), indicating that the subjects with any urinary findings were highly exposed to Cd. In addition, the rates of increased urinary β2-MG were more than 25% in all groups, except for the male glucosuria group. Particularly in women, the β2-MG positive rate was high (36.5%), even in the glucosuria group. These results suggest that not only groups with both glucosuria and proteinuria but also groups with only glucosuria or proteinuria include subjects with renal tubular dysfunction induced by Cd.

Table 1. Age distribution of prevalence of urinary findings: positive cases, smokers and cases with a hypertensive history.

Sex	Age	No (%)	Pro(+)/ Glu(+)(%)	Smoking (%)	HT (%)	No (%)	Pro(+)/ Glu(+)(%)	Smoking (%)	HT (%)
		Control				Exposed			
Men	50–59	383 (41.4)	67 (17.5)	267 (69.7)	80 (20.9)	1443 (42.9)	283 (19.6)	1028 (71.4)	242 (16.8)
	60–69	318 (34.3)	66 (20.8)	186 (58.7)	79 (24.8)	1106 (32.9)	247 (22.3)	674 (61.2)	263 (23.8)
	70–79	179 (19.3)	36 (20.1)	85 (47.5)	60 (33.5)	623 (18.5)	197 (31.6)	319 (51.4)	178 (28.6)
	≥80	46 (5.0)	9 (19.6)	16 (34.8)	10 (21.7)	191 (5.7)	80 (41.9)	56 (29.8)	53 (27.7)
	Total	926 (100.0)	178 (19.2)	554 (59.9)	229 (24.7)	3363 (100.0)	807 (24.0)	2077 (62.0)	736 (21.9)
Women	50–59	474 (40.4)	37 (7.8)	9 (1.9)	97 (20.5)	1613 (40.5)	152 (9.4)	94 (5.8)	254 (15.7)
	60–69	372 (31.7)	41 (11.0)	8 (2.2)	102 (27.4)	1289 (32.3)	237 (18.4)	119 (9.3)	308 (23.9)
	70–79	231 (19.7)	37 (16.0)	5 (2.2)	69 (29.9)	883 (22.2)	304 (34.4)	80 (9.1)	231 (26.2)
	≥80	95 (8.1)	18 (18.9)	6 (6.4)	28 (29.5)	200 (5.0)	108 (54.0)	19 (9.6)	54 (27.0)
	Total	1172 (100.0)	133 (11.3)	28 (2.4)	296 (25.3)	3985 (100.0)	801 (20.1)	312 (7.9)	847 (21.3)

Note, missing 1 control man, 13 exposed men, 4 control women and 11 exposed women from the smoking group, Pro(+)/Glu(+): proteinuria, glucosuria, and glucoproteinuria, No: number of subjects, HT: hypertension history.

Table 2. Median and 90th and 95th percentiles of urinary Cd and positive rate (%) of urinary β2-microglobuline (≥1 mg/dL) in groups with different urinary findings.

Sex	Urinary Findings	N	Urinary Cd (µg/L)			Urinary β2-MG(+)	
			Median	90th per.	95th per.	N	%
Men	Glu(+) and Pro(−)	495	9.1	25.0	38.0	62	12.5
	Pro(+) and Glu(−)	149	10.0	27.0	45.0	40	26.8
	Pro(+) and Glu(+)	159	12.0	22.0	28.9	120	75.5
	Pro(+)/Glu(+)	803	10.0	24.0	34.9	222	27.6
Women	Glu(+) and Pro(−)	395	8.5	20.0	28.0	145	36.7
	Pro(+) and Glu(−)	136	8.5	19.2	37.7	34	25.0
	Pro(+) and Glu(+)	267	7.7	17.9	22.0	242	90.6
	Pro(+)/Glu(+)	793	8.3	19	22.9	421	53.1

Note, N: number of subjects, per.: percentile, β2-MG(+): beta2-microglobuline positive (≥1 mg/dL), Glu(+): glucosuria, Pro(+): proteinuria, Pro(+)/Glu(+): proteinuria, glucosuria, and glucoproteinuria.

2.2. Follow-Up Survey

After obtaining permission to use family registry records for scientific purposes from the regional Legal Affairs Bureau in June of 2005, we collected the registry records of all subjects from each city office and determined their survival status (alive or dead) as of 30 November 2005. Dates and causes of death were determined from vital statistics data after receiving permission from the Ministry of Health and Labor to use vital statistics for research purposes on 12 August 2009. One hundred and sixty-six subjects (1.7%; 133 exposed and 33 control subjects) were excluded because their status could not be ascertained. A total of 5351 deaths were recorded in the exposed and control cohorts, and causes of death for 5276 cases (98.5%) were determined from the records. Individual causes of death were classified according to the Ninth Revised International Classification of Diseases (ICD 9) in the 1979–1994 survey and ICD 10 in the present survey.

2.3. Mortality Analysis

To determine the survival period of each subject, the date of the original health survey was considered the starting point, and 30 November 2005 was considered the end of the follow-up period. A total of 232 subjects, including 157 cases with unknown life status and 75 cases with unknown causes of death, were excluded from the analysis. A mortality risk (hazard) analysis was conducted after adjustment for age, smoking status during the original survey period, and history of hypertension, using the Cox proportional hazards model stratified by sex. Hazard rates were compared between exposed and control subjects, among exposed subjects with different exposure levels, and between subjects with and without urinary findings (glucosuria or proteinuria) during the original survey period. The analyses were performed using SPSS software (Version 21.0, IBM, Armonk, NY, USA, 2012). $p < 0.05$ was considered statistically significant.

3. Results

3.1. Dose–Effect Relationships between Cd Exposure Levels and Prevalence of Proteinuria and/or Glucosuria

We used glucosuria and proteinuria as markers of renal damage induced by Cd, although glucoproteinuria (combined glucosuria and proteinuria) is commonly used as the marker in epidemiological surveys. Therefore, before the mortality analysis, we assessed the associations between these markers and Cd exposure levels in both sexes using logistic regression after adjusting for age, smoking status, and history of hypertension (Table 3). The adjusted odds ratios of having urinary findings of proteinuria, glucosuria, and glucoproteinuria were significantly higher in the Cd-exposed cohort than in the controls for both sexes. Also, the adjusted odds ratios were significantly higher in men in the high-exposure group than in controls. In women, the odds ratio of having urinary findings of proteinuria, glucosuria, and glucoproteinuria increased as the exposure level increased, with significantly higher odds ratios in the mild-, moderate-, and high-exposure groups. These findings suggest that proteinuria, glucosuria, and glucoproteinuria are biomarkers of Cd effects on renal function in both sexes, but the association was higher in women.

Table 3. Prevalence of renal damage indicated by proteinuria, glucosuria, and glucoproteinuria and Cd exposure.

Analysis	Sex	Exposure	No	Pro(+)/Glu(+) (%)	Odds Ratio	(95%CI)	*p*
Model 1	Men	Control	926	178 (19.2)	1.00		
		Exposed	3363	807 (24.0)	1.32	(1.10, 1.59)	0.003
Model 2	Men	Controls	926	178 (19.2)	1.00		
		Non/Boderline	1040	186 (17.9)	0.92	(0.74, 1.16)	0.501
		Mild	1231	275 (22.3)	1.21	(0.98, 1.49)	0.085
		Moderate	515	115 (22.3)	1.18	(0.90, 1.53)	0.236
		High	577	231 (40.0)	2.83	(2.24, 3.59)	0.000
Model 1	Women	Control	1172	133 (11.3)	1.00		
		Exposed	3985	801 (20.1)	2.21	(1.80, 2.72)	0.000
Model 2	Women	Controls	1172	133 (11.3)	1.00		
		Non/Boderline	1228	148 (12.1)	1.13	(0.87, 1.47)	0.347
		Mild	1508	258 (17.1)	1.82	(1.44, 2.30)	0.000
		Moderate	580	129 (22.2)	2.67	(2.02, 3.54)	0.000
		High	669	266 (39.8)	6.20	(4.81, 7.99)	0.000

Note, covariates of logistic analysis: age, HT (hypertension) history, smoking, No: number of subjects, Pro(+)/Glu(+): proteinuria, glucosuria, and glucoproteinuria, CI: confident interval, *p*: *p*-value.

3.2. Comparisons of Cancer Mortality Among Areas with Different Cd Exposure Levels

Table 4 displays the adjusted hazard ratios for mortality from all and from specific malignant neoplasms of all subjects in the Cd-polluted and control areas. No significant increase in the adjusted hazard ratio was found for deaths from malignant neoplasms in either sex. In men, the highest adjusted hazard ratio in the polluted area was observed for deaths from esophageal cancer (1.78), but the increase was not statistically significant. There was no significant difference in mortality from any type of cancer between men in the exposed and control areas. In women, however, the adjusted hazard ratio for colorectal cancer deaths was significantly higher in exposed subjects than in controls, while the hazard ratio for lung cancer deaths was lower with borderline significance.

Table 4. Mortality risk ratios for malignant neoplasms of Cd-exposed subjects compared with controls.

Sex	Cancer	D	HR	D	HR	95%CI
		Controls		Exposed		
		(No = 926)		(No = 3363)		
Men	Total	174	1.00	653	0.98	0.8, 1.2
	Esophagus	4	1.00	27	1.78	0.6, 5.1
	Stomach	57	1.00	180	0.83	0.6, 1.1
	Colon, rectum	16	1.00	64	1.05	0.6, 1.8
	Liver	16	1.00	58	0.96	0.6, 1.7
	Gallbladder	7	1.00	25	0.94	0.4, 2.2
	Pancreas	9	1.00	41	1.15	0.6, 2.4
	Lung	37	1.00	143	1.01	0.7, 1.5
	Prostate	5	1.00	21	1.05	0.4, 2.8
	Bladder	3	1.00	13	1.09	0.3, 3.8
	Kidney/urinal tract	2	1.00	8	1.09	0.2, 5.1
	(Kidney)	0	-	0	-	-
	Lymph/blood forming organs	7	1.00	31	1.15	0.5, 2.6
		(No = 1172)		(No = 3985)		
Women	Total	114	1.00	437	1.05	0.9, 1.3
	Esophagus	0	-	4	-	-
	Stomach	33	1.00	98	0.83	0.6, 1.2
	Colon, rectum	8	1.00	66	2.28	1.1, 4.8 *
	Liver	5	1.00	27	1.43	0.5, 3.7
	Gallbladder	8	1.00	46	1.62	0.8, 3.4
	Pancreas	14	1.00	34	0.67	0.4, 1.3
	Lung	19	1.00	44	0.63	0.4, 1.1
	Breast	4	1.00	14	0.92	0.3, 2.8
	Uterus	3	1.00	15	1.44	0.4, 5.0
	Ovary	2	1.00	9	1.31	0.3, 6.1
	Bladder	2	1.00	4	0.62	0.1, 3.4
	Kidney and urinal tract	1	1.00	8	2.16	0.3, 17.4
	(Kidney)	0	-	4	-	-
	Lymph/blood-forming organs	5	1.00	27	1.39	0.5, 3.6

Note, No: number of subjects, D: number of deaths, HR: hazard ratio, CI: confident interval, *: $p < 0.05$.

To investigate the relationship between Cd exposure levels and cancer mortality, we divided the subjects living in the Cd-polluted area into the four groups with different exposure levels (none/borderline, mild, moderate, and high exposure) reported by Kawano (1996) [13] and analyzed the hazard ratios in the mild-, moderate-, and high-exposure groups compared with those in the none/borderline-exposure group after adjusting for effect-modifying factors (Tables 5 and 6). Men in the high-exposure group had a lower adjusted hazard ratio for deaths from all malignant neoplasms (borderline significance) and for lung cancer deaths. The adjusted hazard ratios for stomach cancer deaths were significantly higher in the mild- and high-exposure groups, and higher (borderline significance) in the moderate-exposure group (Table 5). In women, the only significantly lower hazard ratio was observed for lung cancer deaths in the moderate-exposure group (Table 6).

Table 5. Mortality risk ratios for cancer in three exposed male groups compared with men living in the areas with borderline exposure in the Cd-exposed Jinzu River basin.

Mortality	D	HR	D	HR	95%CI	D	HR	95%CI	D	HR	95%CI
Exposure Levels	Non/Border		Mild			Moderate			High		
No	1040		1231			515			577		
Total	202	1.00	251	1.02	0.84, 1.23	107	1.12	0.88, 1.42	93	0.79	0.61, 1.01 #
Esophagus	13	1.00	9	0.62	0.26, 1.48	4	0.53	0.15, 1.90	1	0.16	0.02, 1.23
Stomach	39	1.00	76	1.58	1.06, 2.36	29	1.62	0.99, 2.64	36	1.61	1.00, 2.57 *
Colon, rectum	22	1.00	24	0.90	0.50, 1.62	10	1.02	0.48, 2.17	8	0.61	0.26, 1.45
Liver	20	1.00	20	0.82	0.44, 1.52	5	0.42	0.14, 1.23	13	1.08	0.52, 2.21
Gallbladder	8	1.00	6	0.58	0.20, 1.68	7	1.77	0.64, 4.90	4	0.79	0.24, 2.63
Pancreas	12	1.00	17	1.11	0.53, 2.32	7	1.19	0.47, 3.02	5	0.54	0.17, 1.67
Lung	50	1.00	58	0.96	0.65, 1.42	21	0.91	0.54, 1.52	14	0.53	0.29, 0.96 *
Prostate	7	1.00	6	0.79	0.25, 2.46	4	1.30	0.37, 4.63	4	0.79	0.20, 3.19
Bladder	5	1.00	4	0.63	0.17, 2.36	3	1.18	0.28, 4.95	1	0.34	0.04, 2.90
Kidney/urinal tract	2	1.00	3	2.22	0.23, 21.3	2	3.79	0.34, 41.9	1	1.56	0.10, 25.2
(Kidney)	0	-	0	-	-	0	-	-	0	-	-
Lymph/blood-forming organs	10	1.00	12	0.97	0.42, 2.25	5	1.00	0.34, 2.94	4	0.67	0.21, 2.16

Note, No: number of subjects, D: number of deaths, HR: hazard ratio, CI: confident interval, #: $p < 0.1$, *: $p < 0.05$.

Table 6. Mortality risk ratios for cancer in three exposed female groups compared with women living in the areas with borderline exposure in the Cd-exposed Jinzu River basin.

Mortality	D	HR	D	HR	95%CI	D	HR	95%CI	D	HR	95%CI
Exposure levels	Non/Border		Mild			Moderate			High		
No	1228		1508			580			669		
Total	144	1.00	167	0.98	0.77, 1.23	56	0.84	0.61, 1.16	70	0.99	0.74, 1.34
Esophagus	1	1.00	1	-	-	1	-	-	1	-	-
Stomach	32	1.00	33	0.91	0.55, 1.51	9	0.66	0.31, 1.40	24	1.50	0.87, 2.62
Colon, rectum	24	1.00	21	0.72	0.40, 1.30	10	0.91	0.43, 1.92	11	0.89	0.43, 1.84
Liver	7	1.00	13	1.90	0.67, 5.41	4	1.16	0.28, 4.85	3	0.73	0.14, 3.75
Gallbladder	11	1.00	21	1.43	0.64, 3.20	9	2.01	0.80, 5.07	5	1.07	0.36, 3.21
Pancreas	12	1.00	12	0.88	0.39, 2.01	4	0.59	0.17, 2.13	6	1.10	0.40, 2.99
Lung	19	1.00	19	0.90	0.47, 1.72	1	0.12	0.02, 0.90 *	5	0.48	0.16, 1.43
Breast	5	1.00	3	0.56	0.12, 2.49	5	2.32	0.62, 8.65	1	0.45	0.05, 4.02
Uterus	5	1.00	6	1.25	0.30, 5.25	1	0.66	0.07, 6.35	3	1.83	0.37, 9.14
Ovary	5	1.00	2	0.30	0.06, 1.53	1	0.40	0.05, 3.39	1	0.39	0.05, 3.36
Bladder	1	1.00	3	2.35	0.24, 22.8	0	-	-	0	-	-
Kidney/urinal tract	1	1.00	4	3.03	0.34, 27.2	2	4.10	0.37, 45.3	1	1.97	0.12, 31.7
(Kidney)	1	1.00	2	1.58	0.14, 1.75	0	-	-	1	2.30	0.14, 38.2
Lymph/blood-forming organs	9	1.00	8	0.70	0.27, 1.83	5	0.91	0.28, 2.97	5	1.07	0.36, 3.20

Note, No: number of subjects, D: number of deaths, HR: hazard ratio, CI: confident interval, *: $p < 0.05$.

3.3. Comparisons of Cancer Mortality between Exposed Residents with and without Cd-Induced Renal Damage

Renal effects, indicated by proteinuria, glucosuria, and glucoproteinuria, were associated with higher exposure to Cd (Table 3), but it is possible that the glucosuria group included patients with Type 2 diabetes who are at risk of developing malignancies. Therefore, we excluded 10 subjects (five men and five women) who showed increased fasting blood sugar more than 125 (mg/dL) in the second step of the baseline health screening test from the subjects for the mortality analysis. Then, we calculated the hazard ratios for subjects with urinary findings after adjustment, compared with ratios in subjects without urinary findings in the Cd-polluted area (Table 7).

In men, the lower adjusted hazard ratio for deaths from total malignant neoplasms was not significant, but the hazard ratio for lung cancer deaths was significantly lower in men with urinary findings. In contrast, the adjusted hazard ratio for cancer deaths of the kidneys and urinal tract was

2.39, but the increase in the hazard ratio was not significant in men. In women, adjusted hazard ratios for deaths from total malignant neoplasms and kidney and urinary tract cancers, particularly kidney cancer, were significantly higher in the Cd-exposed subjects with urinary findings. The hazard ratio for deaths from uterine cancer was also significantly higher in the subjects with urinary findings. The adjusted hazard ratio from pancreas cancer was higher (1.99), but its significance was borderline ($p = 0.093$).

Table 7. Adjusted hazard ratios for cancer in subjects with proteinuria, glucosuria, and glucoproteinuria, compared with subjects without urinary findings living in the exposed area.

Sex	Cancer	D	HR	D	HR	95%CI
		No Findings		Proteinuria and/or Glucosuria		
		(No = 2556)		(No = 802)		
	Total	522	1.00	130	0.87	0.71, 1.06
	Esophagus	20	1.00	7	1.13	0.45, 2.86
	Stomach	142	1.00	38	0.95	0.66, 1.37
	Colon, rectum	53	1.00	10	0.73	0.37, 1.43
	Liver	42	1.00	16	1.44	0.80, 2.59
	Gall bladder	21	1.00	4	0.68	0.23, 1.98
Men	Pancreas	33	1.00	8	0.91	0.42, 1.98
	Lung	121	1.00	22	0.62	0.39, 0.98 *
	Prostate	13	1.00	8	1.59	0.60, 4.25
	Bladder	11	1.00	2	0.59	0.13, 2.71
	Kidney/urinal tract	5	1.00	3	2.39	0.52, 10.9
	(Kidney)	0	1.00	0	-	-
	Lymph/blood-forming organs	24	1.00	7	1.02	0.44, 2.39
		(No = 3184)		(No = 796)		
	Total	339	1.00	97	1.49	1.17, 1.89 *
	Esophagus	2	1.00	2	1.82	0.15, 21.7
	Stomach	77	1.00	21	1.19	0.71, 1.98
	Colon, rectum	55	1.00	11	0.87	0.44, 1.70
	Liver	24	1.00	3	0.60	0.14, 2.61
	Gallbladder	36	1.00	10	1.75	0.84, 3.65
	Pancreas	25	1.00	9	1.99	0.89, 4.44 #
Women	Lung	35	1.00	9	1.40	0.65, 3.00
	Breast	11	1.00	3	2.21	0.59, 8.28
	Uterus	10	1.00	5	3.85	1.16, 12.8 *
	Ovary	7	1.00	2	1.58	0.31, 8.02
	Bladder	4	1.00	0	-	-
	Kidney and urinal tract	3	1.00	5	10.1	2.29, 44.5 *
	(Kidney)	2	1.00	2	7.71	1.05, 56.8 *
	Lymph/blood-forming organs	20	1.00	6	1.98	0.77, 5.09

Note, No (urinary) findings: nether proteinuria nor glucosuria, No: number of subjects, D: number of deaths, HR: hazard ratio, CI: confident interval, #: $p < 0.1$, *: $p < 0.05$.

4. Discussion

4.1. Cd Exposure and Cancer Mortality

In Cd-exposed women, the hazard ratio for deaths from colorectal cancer differed significantly from that of controls, although no dose–response relationship was found between Cd exposure level and hazard ratio for deaths from any malignant neoplasms. This lack of relationship suggests that women living in the Cd-polluted Jinzu River basin area are not at a greater risk of dying of cancer.

In men, the adjusted hazard ratios for stomach cancer deaths were significantly higher for subjects exposed to mild, moderate, and high levels of Cd. No dose–response relationship, however, was found between Cd exposure levels and risk of stomach cancer deaths. The hazard ratio for lung cancer deaths was significantly lower in men exposed to high levels of Cd who had a historical record of proteinuria, glucosuria, and glucoproteinuria, suggesting a possible inverse association between Cd exposure or Cd-induced renal damage and lung cancer.

In our previous 22-year follow-up study in residents of another Cd-polluted area, the Kakehashi River basin in Japan, no association was found between Cd exposure levels, as indicated by urinary Cd content, and deaths from stomach and lung cancer in men. In women, however, hazard ratios for lung cancer deaths were significantly lower in the subjects with urinary Cd $\geq$10 μg/g Cr than in subjects with Cd <10 μg/g Cr [15]. These findings suggest an inverse association between Cd exposure and lung cancer mortality, although the findings differed by sex.

Two meta-analyses of lung cancer risks associated with Cd exposure have been published, but their findings were inconsistent. Nawrot et al. (2015) analyzed three cohort studies and reported a significantly higher risk for lung cancer with environmental Cd exposure [16]. Chen et al. (2016) included environmental and occupational exposures, and could not find significant associations between Cd exposure and an increased risk of lung cancer [17].

4.2. Cd-Induced Renal Damage and Cancer Mortality

Renal tubular dysfunction is a characteristic symptom of chronic Cd poisoning and is indicated by glucoproteinuria (both proteinuria and glucosuria). We previously found increased cancer mortality in women with glucoproteinuria in the same cohort in the Jinzu river basin (Maruzeni et al. 2014). In the present analysis, we used proteinuria and/or glucosuria as a marker of Cd-induced renal effects. Their Cd exposure levels were remarkably high, with the 95th percentile urinary Cd $\geq$20 μg/L, suggesting Cd exposure from environmental pollution. Because the 95th percentile of urinary Cd was reported to be 3.50 μg/L in smokers aged $\geq$50 years old [18], this suggested that smoking alone, a major cause of Cd exposure in the general population, cannot increase Cd exposure levels. The prevalence of proteinuria, glucosuria, and glucoproteinuria increased as Cd exposure levels increased, suggesting that either proteinuria or glucosuria can also serve as indicators of Cd-induced renal damage. However, type 2 diabetes is a risk factor for developing cancers, and subjects with only glucosuria might include patients with type 2 diabetes. Therefore, we deleted subjects at a risk of diabetes who showed fasted blood sugar $\geq$125 (mg/dL) from the subjects with proteinuria, glucosuria, and glucoproteinuria and analyzed their mortality risk ratios compared with the subjects without urinary findings.

In the present analysis, the mortality risk from malignant neoplasms in women was significantly higher in Cd-exposed subjects who had a history of urinary findings of proteinuria, glucosuria, and glucoproteinuria. Deaths from kidney and urinary tract cancers, including renal cancer, uterine cancer, and pancreatic cancer, were observed in this cohort. These findings were not detected in our previous analysis, which used glucoproteinuria as a biomarker of renal effects [10]. The different results between our present and previous analysis suggest that in Cd-polluted areas, women who present with proteinuria, glucosuria, and glucoproteinuria may be at higher risk of death from kidney and urinary tract cancers, particularly renal cancer. In addition, we analyzed adjusted hazard ratios for deaths from renal and urinary tract cancer in a model including two factors—renal damage and exposure levels—in women. We found that only the renal damage factor was significantly associated with mortality risk. This finding suggests that Cd-induced renal damage may be a risk factor for death from renal cancer in women living in the Cd-polluted area.

Il'yasova and Schwartz (2005) performed a systematic review of studies on Cd exposure and renal cancer, and reported an increased risk in large-scale epidemiological studies with data from four countries [9]. Cd content in the kidney cortex is not an effective marker of Cd accumulation following renal cellular damage [19], and other markers should be used to evaluate exposure levels in cancer patients. Therefore, an association between Cd exposure and renal cancer could not be concluded from the findings in these clinical studies.

We reported an increased risk of mortality from uterine cancer in women with renal damage in our previous study [10], confirming the positive association between Cd exposure and death from uterine cancer observed in the present analysis. In the USA, increased mortality from uterine cancer was reported to be associated with increasing urinary Cd levels in a study using data from the Third National Health and Nutrition Examination Survey (NHANES III) [8]. These results suggest an increased mortality risk from uterine cancer is associated with increasing Cd exposure.

In the same mortality study by Adams et al. (2012), significantly increased mortality from pancreatic cancer associated with urinary Cd was reported in men [8]. In the meta-analysis of cohorts with highly Cd-exposed workers, Schwartz and Reis (2000) indicated there was an increased risk of pancreatic cancer with borderline significance (standardized mortality ratio = 166; p = 0.059) in both sexes [20]. In the present study, increased mortality from prostatic cancer was also of borderline significance, but only in women. A further follow-up study may be required to determine whether Cd exposure influences the development of pancreatic cancers.

In men with historical records of proteinuria, glucosuria, and glucoproteinuria, the adjusted hazard ratios for mortality from kidney and urinary tract cancers were higher, with borderline significance. However, no death from kidney cancer was found in men with urinary findings; thus, the effects of Cd-induced renal damage on renal cancer mortality may be limited in men. In addition, the adjusted hazard ratio for prostate cancer deaths was higher, but its increase was not significant. An increased prostate cancer risk has been reported in several follow-up surveys in industrial workers in Europe [3,4,21]. In our study in the Kakehashi River basin as well as in the present study, no association between renal damage and prostate cancer deaths was found [22]. This lack of association may be because the European studies measured the incidence of prostate cancer, while we only measured mortality and because their cohorts had occupational exposure, while the exposure in our study was environmental.

Mortality from lung cancer was significantly lower in Cd-exposed men with a historical record of urinary findings. Since a lower risk of mortality from lung cancer was found in the dose–response analysis, we performed additional calculations with an exposure factor and a renal-effect factor and found that both factors were independently associated, with borderline significance, with decreased mortality from lung cancer in men. In our previous study [10], we did not find differences in risk of mortality from lung cancer in Cd-exposed men with glucoproteinuria. Therefore, we cannot conclude that Cd-induced renal damage decreases the mortality risk from lung cancer.

4.3. Limitations

Our present study has several limitations that should be considered. Follow-up data from the subjects included mortality from cancers but not their incidence. Therefore, we could not estimate cancer risk, particularly for cancers with high survival rates, such as prostate, uterine, and breast cancers. Stomach cancer can be found and treated at early stages, suggesting that its incidence may be a more important indicator of risk than mortality. For renal cancer, however, mortality may be an adequate risk indicator, because renal cancer is difficult to diagnose at an early stage, and most cases are fatal.

Another limitation is that environmental Cd exposure levels were based on the contribution of contaminated Jinzu River water [13], and not biological markers of Cd body burden, such as urinary levels in each subject, which are more closely related to renal effects. In addition, we used historical exposure levels from the baseline survey, and did not measure Cd exposure during the follow-up period. Cd body burden, however, may not have significantly increased during the follow-up period, because ingestion of Cd-polluted rice and river water was prohibited in the Jinzu River basin, and residents are presumed to have had no additional Cd exposure during that period.

We used proteinuria, glucosuria, and glucoproteinuria as markers of renal damage in this analysis, but these urinary findings are not specific to renal tubular dysfunction. We first analyzed the dose–response relationship between Cd exposure levels and the prevalence of proteinuria and/or glucosuria to confirm that this marker indicated renal effects associated with Cd exposure. Urinary Cd levels and urinary β2-MG positive rates of the subjects with proteinuria, glucosuria, and glucoproteinuria were high enough to develop renal effects. Moreover, the subjects who showed glucosuria (without proteinuria) and diabetic elevation of blood sugar in the second step of the baseline health screening test were excluded from the mortality analysis to avoid confounding by the presence of type 2 diabetes. After these modifications, we believe that proteinuria, glucosuria, and glucoproteinuria can be used as markers of renal damage in these subjects living in a heavily contaminated area to Cd in Japan.

5. Conclusions

Our findings suggest that women with renal damage associated with high Cd exposure levels are at risk of mortality from malignant neoplasms, including renal and uterine cancers. Mortality risk from pancreatic cancer may be increased in women with Cd-induced renal damage, but more studies are required to confirm this hypothesis.

Acknowledgments: This work was supported by grant for aids from Agency of Environment for Health effects due to Heavy Metal exposure 2009–2012 and 2013–2015, 2016–2018. We thank Dean Meyer, ELS from Edanz Group (www.edanzediting.com/ac) for editing a draft of this manuscript.

Author Contributions: M.N., H.N. and T.K. conceived and designed the experiments; Y.S., K.N. and M.N. analyzed the data; M.S. and M.I. contributed data linkage and making data base; M.N. and H.N. wrote the paper.

Conflicts of Interest: The authors declare no conflict of interest. The founding sponsors had no role in the design of the study; in the collection, analyses, or interpretation of data; in the writing of the manuscript, and in the decision to publish the results.

References

1. International Agency for Research on Cancer. Cadmium and Cadmium Compounds. In *IARC Monograph*; International Agency for Research on Cancer: Lyon, France, 2012; Volume 100C, pp. 121–145.
2. Nawrot, T.; Plusquin, M.; Hogervorst, J.; Roels, H.A.; Celis, H.; Thijs, L.; Vangronsveld, J.; Van Hecke, E.; Staessen, J.A. Environmental exposure to cadmium and risk of cancer: A prospective population-based study. *Lancet Oncol.* **2006**, *7*, 119–126. [CrossRef]
3. Elinder, C.G.; Kjellstrom, T.; Hogstedt, C.; Andersson, K.; Spang, G. Cancer mortality of cadmium workers. *Br. J. Ind. Med.* **1985**, *42*, 651–655. [CrossRef] [PubMed]
4. Armstrong, B.G.; Kazantzis, G. Prostatic cancer and chronic respiratory and renal disease in British cadmium workers: A case control study. *Br. J. Ind. Med.* **1985**, *42*, 540–545. [CrossRef] [PubMed]
5. McElroy, J.A.; Shafer, M.M.; Trenthan-Dietz, A.; Hampton, J.M.; Newcomb, P.A. Cadmium exposure and breast cancer risk. *J. Natl. Cancer Inst.* **2006**, *98*, 869–873. [CrossRef] [PubMed]
6. Julin, B.; Wolk, A.; Bergkvist, L.; Bottai, M.; Akesson, A. Dietary cadmium exposure and risk of postmenopausal breast cancer: A population-based prospective cohort study. *Cancer Res.* **2012**, *72*, 1459–1466. [CrossRef] [PubMed]
7. Akesson, A.; Julin, B.; Wolk, A. Long-term dietary cadmium intake and postmenopausal endometrial cancer incidence: A population-based prospective cohort study. *Cancer Res.* **2008**, *68*, 6435–6441. [CrossRef] [PubMed]
8. Adams, S.V.; Passarelli, M.N.; Newcomb, P.A. Cadmium exposure and cancer mortality in the Third National Health and Nutrition Examination Survey cohort. *Occup. Environ. Med.* **2012**, *69*, 153–156. [CrossRef] [PubMed]
9. Ilyasova, D.; Schwartz, G.G. Cadmium and renal cancer. *Toxicol. Appl. Pharmacol.* **2005**, *207*, 179–186. [CrossRef] [PubMed]
10. Maruzeni, S.; Nishijo, M.; Nakamura, K.; Morikawa, Y.; Sakurai, M.; Nakashima, M.; Kido, T.; Okamoto, R.; Nogawa, K.; Suwazono, Y.; et al. Mortality and causes of deaths of inhabitants with renal dysfunction induced by cadmium exposure of the polluted Jinzu River basin, Toyama, Japan; a 26-year follow-up. *Environ. Health* **2014**, *13*, 18. [CrossRef] [PubMed]

11. Friberg, L. The itai-itai disease. In *Cadmium in the Environment*; Friberg, L., Piscator, M., Nordberg, G., Eds.; CRC Press, The Chemical Rubber Co.: Cleveland, OH, USA, 1971; pp. 111–114.

12. Nogawa, K.; Kido, T. Itai-itai disease and health effects of cadmium. In *Toxicology of Metals*; Chang, L.W., Ed.; CRC Press: New York, NY, USA, 1996; pp. 353–369.

13. Kawano, S. Epidemiology of itai-itai disease. *J. Hokuriku Public Health* **1996**, *23*, 45–52.

14. Japan Public Health Association Cadmium Research Committee. Report on studies of health effects of cadmium. Appendices; Health survey in inhabitants living in cadmium polluted areas. *Kankyo Hoken Rep.* **1989**, *56*, 69–345.

15. Li, Q.; Nishijo, M.; Nakagawa, H.; Morikawa, Y.; Sakurai, M.; Nakamura, K.; Kido, T.; Nogawa, K.; Dai, M. Dose-response relationship between urinary cadmium and mortality in habitants of a cadmium-polluted area: A 22-year follow-up study in Japan. *Chin. Med. J.* **2011**, *124*, 3504–3509. [PubMed]

16. Nawrot, T.S.; Martens, D.S.; Hara, A.; Plusquin, M.; Vangronsveld, J.; Roels, H.A.; Staessen, J.A. Association of total cancer and lung cancer with environmental exposure to cadmium: The meta-analytical evidence. *Cancer Causes Control* **2015**, *26*, 1281–1288. [CrossRef] [PubMed]

17. Chen, C.; Xun, P.; Nishijo, M.; He, K. Cadmium exposure and risk of lung cancer: A meta-analysis of cohort and case-control studies among general and occupational populations. *J. Expo. Sci. Environ. Epidemiol.* **2016**, *26*, 437–444. [CrossRef] [PubMed]

18. Richter, P.; Faroon, O.; Pappas, R.S. Cadmium and cadmium/zinc ratios and tobacco-related morbidities. *Int. J. Environ. Res. Public Health* **2017**, *14*, 1154. [CrossRef] [PubMed]

19. Honda, R.; Nogawa, K. Cadmium, zinc and copper relationships in kidney and liver of humans exposed to environmental cadmium. *Arch. Toxicol.* **1987**, *59*, 437–442. [CrossRef] [PubMed]

20. Schwartz, G.G.; Reis, I.M. Is cadmium a cause of human pancreatic cancer? *Cancer Epidemiol. Biomark. Prev.* **2000**, *9*, 139–145.

21. Sorahan, T.; Esmen, N.A. Lung cancer mortality in UK nickel-cadmium battery workers, 1947–2000. *Occup. Environ. Med.* **2004**, *61*, 108–116. [CrossRef] [PubMed]

22. Nishijo, M.; Morikawa, Y.; Nakagawa, H.; Tawara, K.; Miura, K.; Kido, T.; Ikawa, A.; Kobayashi, E.; Nogawa, K. Causes of death and renal tubular dysfunction in residents exposed to cadmium in the environment. *Occup. Environ. Med.* **2006**, *63*, 545–550. [CrossRef] [PubMed]

Brief Report

Effects of Cadmium Exposure on Age of Menarche and Menopause

Xiao Chen [1], Guoying Zhu [2] and Taiyi Jin [3,*]

[1]   Department of Nephrology, Zhongshan Hospital, Fudan University, 180 Fenglin Road, Shanghai 200032, China; chxwin@163.com
[2]   Institute of Radiation Medicine, Fudan University, 2094 Xietu Road, Shanghai 200032, China; zhugy@shmu.edu.cn
[3]   Department of Occupational Medicine, School of Public Health, Fudan University, 150 Dongan Road, Shanghai 200032, China
*   Correspondence: tyjin@shmu.edu.cn; Tel.: +86-021-5423-7625

Received: 7 December 2017; Accepted: 26 December 2017; Published: 27 December 2017

Abstract: Cadmium exposure can cause several adverse health effects. Animal studies have also shown that cadmium exposure can affect menarche or menopause. However, data is limited in humans. We conducted a retrospective study to assess whether cadmium exposure was associated with different ages of menarche and menopause in a Chinese population. A total of 429 women living in control ($n = 137$) and two cadmium-polluted areas ($n = 292$) were included in this study. A total of 223 and 206 subjects were included in the analysis for menarche and menopause, respectively. The median menarche age of population living in the heavily cadmium-polluted area was significantly younger than those in the control area (14.0 vs. 15.0, $p < 0.01$). Logistic regression showed that the odds ratio (OR) of early occurrence of menarche (<13 years) in the population living in the heavily polluted area and moderately polluted area was 3.7 (95% confidence interval (CI): 1.5–9.7) and 1.3 (95% CI: 0.7–2.6) compared with control, respectively. No significant difference was observed in the age of menopause in the population of these three areas. In conclusion, our data indicated that cadmium exposure may cause early menarche.

Keywords: cadmium; female; menarche; menopause

1. Introduction

Cadmium exposure has been shown to adversely affect the liver, kidney, cardiovascular system, and bones. Cadmium also acts as an endocrine disruptor and affects reproduction system [1]. Animal studies have shown that cadmium exposure can cause an increase of uterine wet weight, endometrial thickness, and endometrial stromal thickness in female rats [2]. Studies also indicated that maternal exposure to cadmium can increase early delivery and lower birth weight [3,4]. However, very little human data were available on the association between environmental level of cadmium exposure and age of menarche and menopause [5]. In the present study, we examined the association of cadmium exposure and menarche age and menopause age in a Chinese population.

2. Materials and Methods

2.1. Study Area and Population

A ChinaCad study was performed during 1997, which aimed to observe the influence of cadmium exposure on renal dysfunction and osteoporosis [6]. The following three areas were included in the present study: Nanbaixiang ('moderate' near Wenzhou, cadmium in rice = 0.51 mg/kg), Jiaoweibao ('heavy' near Wenzhou, mean cadmium concentration in rice = 3.7 mg/kg), and a control area (40 km from Wenzhou, Cd concentration in rice = 0.07 mg/kg). There was a smelter in the heavily polluted area that began production in 1961. The waste water was directly discharged into the river. Residents living in the polluted areas used the polluted river water to irrigate their fields from 1961 to 1995. The cadmium concentration in rice was 3.7 mg/kg in 1997. An area with similar nutrition and socioeconomic factors and low cadmium exposure was selected as the control. All subjects gave their informed consent for inclusion before they participated in the study. The study was conducted in accordance with the Declaration of Helsinki, and the protocol was approved by the Ethics Committee of Fudan University.

A total of 488 women were included in this study. As the pollution started from 1961, we analyzed the association between menarche age and cadmium exposure in subjects <47 years and the association between menopause age and cadmium exposure in people that are 50–80 years old. A total of 429 women was finally included in this study, including 137 women in control area and 292 women in two cadmium polluted areas. A total of 223 subjects was included in menarche study and 206 were included in the menopause study. The information on menarche age and menopause age were obtained through self-reporting by subjects.

2.2. Statistical Analysis

The data was analyzed using SPSS 16.0 (SPSS Inc., Chicago, IL, USA). Quantitative data were shown as means ± standard deviation or standard error and were analyzed by one-way analysis of variance (ANOVA) or analysis of covariance (adjusted with exposure duration). An earlier age of menarche was defined as age of <13 years old, while delayed menopause was defined as menopause age of >51 years old. Logistic regression was used to show the risk of an abnormal menarche age at different levels of cadmium exposure. *p*-values of less than 0.05 were considered to be statistically significant.

3. Results

The median menarche age was 14 (11–16) in the heavily polluted area, 14 (11–18) in the moderately polluted area, and 15 (12–17) in the control area. The mean age of menarche was 13.9 in heavily polluted area, 14.4 in moderately polluted area, and 15.2 in control area (Figure 1), respectively. After adjusting for the exposure duration, the mean age of menarche was 14.0 in the heavily polluted area, 14.4 in the moderately polluted area, and 15.0 in the control area (Figure 2). The menarche age was approximately one year younger for subjects in the heavily polluted area compared with the subjects in the control area. The menarche age of subjects living in the heavily polluted area was significantly younger than those in the control area ($p < 0.001$). An earlier onset of menarche was observed in 6.5%, 12.5%, and 21.1% of subjects living in the control, moderately polluted area, and heavily polluted area, respectively.

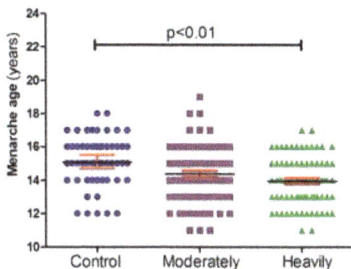

Figure 1. The menarche age of women in control (*n* = 57), moderately (*n* = 89) and heavily (*n* = 77) polluted areas.

Logistic regression further showed that the odds ratio (OR) of early occurrence of menarche in the population in the heavily polluted area was 3.7 (95% confidence interval (CI): 1.5–9.7) compared with those people living in the control area (Figure 3). The population living in moderately polluted area also had a high risk of early menarche, but this was not statistically significant (OR = 1.3, 95% CI: 0.7–2.6).

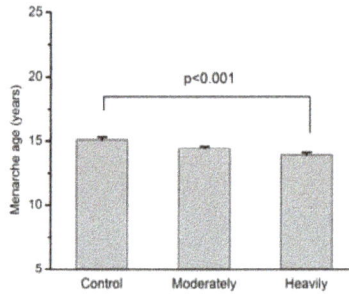

Figure 2. The menarche age of women in control (*n* = 57), moderately (*n* = 89) and heavily (*n* = 77) polluted areas after adjusting with exposure duration.

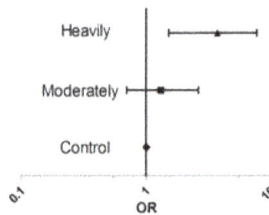

Figure 3. The odds ratio (OR) of early occurrence of menarche and cadmium exposure (control (*n* = 57), moderately (*n* = 89), and heavily (*n* = 77) polluted areas).

The menopausal age was 47.4, 47.5, and 47.0 in the women living in the heavily polluted area, moderately polluted area, and control area, respectively. After adjustment for exposure duration, the menopausal age was 47.3, 47.5, and 47.1, respectively. No significant differences were observed. A delayed menopause was observed in 12.3%, 15.1%, and 17.6% of subjects living in the control, moderately polluted, and heavily polluted areas, respectively.

4. Discussion

In the present study, we showed that high levels of cadmium exposure can induce early menarche. The average menarche age was approximately one year younger in subjects in the heavily polluted area compared with the subjects in the control area.

Previous data showed that the main contributors to dietary cadmium are rice, vegetables, and pork in the Chinese diet [7,8]. For our populations, 80% of dietary cadmium is from rice because the concentrations of cadmium in rice are high, which may be different from other general populations [7]. The dietary pattern of our population is similar to the traditional southern dietary pattern (high intake of rice, vegetables, and pork) [7].

Many factors will affect the timing of pubertal onset, such as nutrition, obesity, and environmental contaminants [9]. Heavy metal exposure is also a factor that may influence changes in the timing of pubertal onset. Previous studies show that lead exposure is associated with later menarche [10,11]. Cadmium exposure results in uterine hyperplasia and early onset of puberty in female rats [1]. However, the effects of cadmium on menarche and menopause in humans have not been fully investigated [12]. The age of menarche is younger than in the past [5]. It has been shown that endocrine disrupting chemicals may play a critical role in this phenomenon [11]. Our data further confirmed that cadmium exposure is associated with the shift in age of menarche. A recent study showed that a higher cadmium body burden (unadjusted urinary cadmium (UCd) > 0.4 µg/L) was less likely to have attained menarche during two years follow-up [13]. The cadmium levels were lower than the exposure in our study. In addition, the study did not show the association between adjusted UCd (shown as µg/g crentinine) and menarche. Moreover, only 12 to 15-year-old girls were included in that study. Further studies are needed. It has been shown that cadmium could be regarded as a potent nonsteroidal estrogen [1]. Cadmium may mimic the effects of estradiol in promoting uterine hyperplasia, development of mammary glands, and earlier menarche [1]. In addition, cadmium can activate membrane-bound estrogen receptors [14] and enhance the activity of estrogen.

The association between cadmium exposure and age at menopause has not been clarified [5,15]. In the present study, we found that there were no significant differences in the timing of menopause between subjects living in the control area and cadmium polluted areas. The prevalence of delayed menopause was slightly increased in cadmium polluted areas compared with that in the control area. Cadmium may mimic the effects of estradiol in delaying the occurrence of menopause.

There are several limitations in our study. First, we could not obtain the internal dose of cadmium, including urinary cadmium (UCd) and blood cadmium (BCd), when the subjects were at menarche or menopause. Second, the logistic model did not adjust for some confounders, such as nutritional status or body weight when the subjects were young. Third, the population size was relatively small. Further study with larger sample sizes should be performed.

In conclusion, our data indicated that cadmium exposure is associated with earlier menarche. Cadmium exposure may slightly delay the age of menopause. Further studies are needed because women are more susceptible to cadmium-induced health effects.

Acknowledgments: This study was funded by National Natural Science foundation of China (No. 81773460, 81102148) and the Natural Science Foundation of Jiangsu Province (No. BK20161609).

Author Contributions: X.C. and T.J. conceived and designed the experiments; X.C. and G.Z. performed the study; X.C. analyzed the data; X.C. wrote the draft; X.C., G.Z., and T.J. revised the paper

Conflicts of Interest: The author declares there is no conflict of interest.

References

1. Johnson, M.D.; Kenney, N.; Stoica, A.; Hilakivi-Clarke, L.; Singh, B.; Chepko, G.; Clarke, R.; Sholler, P.F.; Lirio, A.A.; Foss, C.; et al. Cadmium mimics the in vivo effects of estrogen in the uterus and mammary gland. *Nat. Med.* **2003**, *9*, 1081–1084. [CrossRef] [PubMed]

2. Hofer, N.; Diel, P.; Wittsiepe, J.; Wilhelm, M.; Kluxen, F.M.; Degen, G.H. Dose and route-dependent hormonal activity of the metalloestrogen cadmium in the rat uterus. *Toxicol. Lett.* **2009**, *191*, 123–131. [CrossRef] [PubMed]

3. Nishijo, M.; Nakagawa, H.; Honda, R.; Tanebe, K.; Saito, S.; Teranishi, H.; Tawara, K. Effects of maternal exposure to cadmium on pregnancy outcome and breast milk. *Occup. Environ. Med.* **2002**, *59*, 394–397. [CrossRef] [PubMed]

4. Ronco, A.M.; Urrutia, M.; Montenegro, M.; Llanos, M.N. Cadmium exposure during pregnancy reduces birth weight and increases maternal and foetal glucocorticoids. *Toxicol. Lett.* **2009**, *188*, 186–191. [CrossRef] [PubMed]

5. Pollack, A.Z.; Ranasinghe, S.; Sjaarda, L.A.; Mumford, S.L. Cadmium and Reproductive Health in Women: A Systematic Review of the Epidemiologic Evidence. *Curr. Environ. Health Rep.* **2014**, *1*, 172–184. [CrossRef] [PubMed]

6. Jin, T.; Nordberg, M.; Frech, W.; Dumont, X.; Bernard, A.; Ye, T.T.; Kong, Q.; Wang, Z.; Li, P.; Lundström, N.G.; et al. Cadmium biomonitoring and renal dysfunction among a population environmentally exposed to cadmium from smelting in China (ChinaCad). *Biometals* **2002**, *15*, 397–410. [CrossRef] [PubMed]

7. Shi, Z.; Taylor, A.W.; Riley, M.; Byles, J.; Liu, J.; Noakes, M. Association between dietary patterns, cadmium intake and chronic kidney disease among adults. *Clin. Nutr.* **2017**, *5614*, 31366–31368. [CrossRef] [PubMed]

8.   He, P.; Lu, Y.; Liang, Y.; Chen, B.; Wu, M.; Li, S.; He, G.; Jin, T. Exposure assessment of dietary cadmium: Findings from Shanghainese over 40 years, China. *BMC Public Health* **2013**, *13*, 590. [CrossRef] [PubMed]

9.   Buck Louis, G.M.; Gray, L.E., Jr.; Marcus, M.; Ojeda, S.R.; Pescovitz, O.H.; Witchel, S.F.; Sippell, W.; Abbott, D.H.; Soto, A.; Tyl, R.W.; et al. Environmental factors and puberty timing: Expert panel research needs. *Pediatrics* **2008**, *121*, S192–S207. [CrossRef] [PubMed]

10. Wu, T.; Buck, G.M.; Mendola, P. Blood lead levels and sexual maturation in U.S. girls: The Third National Health and Nutrition Examination Survey, 1988–1994. *Environ. Health Perspect.* **2003**, *111*, 737–741. [CrossRef] [PubMed]

11. Gollenberg, A.L.; Hediger, M.L.; Lee, P.A.; Himes, J.H.; Louis, G.M. Association between lead and cadmium and reproductive hormones in peripubertal U.S. girls. *Environ. Health Perspect.* **2010**, *118*, 1782–1787. [CrossRef] [PubMed]

12. Vahter, M.; Berglund, M.; Akesson, A. Toxic metals and the menopause. *J. Br. Menopause Soc.* **2004**, *10*, 60–64. [CrossRef] [PubMed]

13. Rull, R.P.; Canchola, A.J.; Reynolds, P.; Horn-Ross, P.L. Urinary Cadmium and the Timing of Menarche and Pubertal Development in Girls. *J. Adolesc. Health* **2014**, *54*, S10–S11. [CrossRef]

14. Yu, X.; Filardo, E.J.; Shaikh, Z.A. The membrane estrogen receptor GPR30 mediates cadmium-induced proliferation of breast cancer cells. *Toxicol. Appl. Pharmacol.* **2010**, *245*, 83–90. [CrossRef] [PubMed]

15. Pollack, A.Z.; Louis, G.M.; Chen, Z.; Peterson, C.M.; Sundaram, R.; Croughan, M.S.; Sun, L.; Hediger, M.L.; Stanford, J.B.; Varner, M.W.; et al. Trace elements and endometriosis: The ENDO Study. *Reprod. Toxicol.* **2013**, *42c*, 41–48. [CrossRef] [PubMed]

toxics

MDPI

Article

Intrauterine Exposure to Cadmium Reduces HIF-1 DNA-Binding Ability in Rat Fetal Kidneys

Tania Jacobo-Estrada [1,2] 🔟, Mariana Cardenas-Gonzalez [3] 🔟, Mitzi Paola Santoyo-Sánchez [1], Frank Thevenod [4] and Olivier Barbier [1,*] 🔟

[1]   Departamento de Toxicología, Centro de Investigación y de Estudios Avanzados del Instituto Politécnico Nacional, 07360 Ciudad de México, Mexico; tjacoboe@ipn.mx (T.J.-E.); santoyomitzi@gmail.com (M.P.S.-S.)
[2]   Departamento de Sociedad y Política Ambiental, CIIEMAD, Instituto Politécnico Nacional, 07340 Ciudad de México, Mexico
[3]   Renal Division, Department of Medicine, Brigham and Women's Hospital, Harvard Medical School, Boston, MA 02115, USA; mcardenasgonzalez@bwh.harvard.edu
[4]   Department of Physiology, Pathophysiology and Toxicology and ZBAF (Center for Biomedical Education and Research), Faculty of Health-School of Medicine, Witten/Herdecke University, 58448 Witten, Germany; frank.thevenod@uni-wh.de
*   Correspondence: obarbier@cinvestav.mx; Tel.: +52-1-(55)-5747-3800

Received: 11 June 2018; Accepted: 29 August 2018; Published: 3 September 2018

Abstract: During embryonic development, some hypoxia occurs due to incipient vascularization. Under hypoxic conditions, gene expression is mainly controlled by hypoxia-inducible factor 1 (HIF-1). The activity of this transcription factor can be altered by the exposure to a variety of compounds; among them is cadmium (Cd), a nephrotoxic heavy metal capable of crossing the placenta and reaching fetal kidneys. The goal of the study was to determine Cd effects on HIF-1 on embryonic kidneys. Pregnant Wistar rats were exposed to a mist of isotonic saline solution or $CdCl_2$ (D_{Del} = 1.48 mg Cd/kg/day), from gestational day (GD) 8 to 20. Embryonic kidneys were obtained on GD 21 for RNA and protein extraction. Results show that Cd exposure had no effect on HIF-1α and prolyl hydroxylase 2 protein levels, but it reduced HIF-1 DNA-binding ability, which was confirmed by a decrease in *vascular endothelial growth factor* (*VEGF*) mRNA levels. In contrast, the protein levels of VEGF were not changed, which suggests the activation of additional regulatory mechanisms of VEGF protein expression to ensure proper kidney development. In conclusion, Cd exposure decreases HIF-1-binding activity, posing a risk on renal fetal development.

Keywords: cadmium; embryonic kidneys; HIF-1; intrauterine exposure

1. Introduction

Hypoxia plays an important role in various processes during fetal development, like placentation, angiogenesis, and hematopoiesis [1]. Renal development is also driven by low partial oxygen pressure, which in the rat initiates after the implantation and the apparition of the primitive streak, around gestational day (GD) 8 or 9 [2]. Nephrogenesis and the growth and development of renal vasculature occur simultaneously, which causes a disparity of oxygen demand and supply due to a low degree of vascularization. This generates local low oxygen tension in early developmental stages, so it is believed that hypoxia somehow regulates tissue maturation during these developmental stages [3]. Under hypoxic conditions, gene expression is primarily controlled by the hypoxia-inducible factor 1 (HIF-1). HIF-1 is a transcription factor composed by two subunits: 1α and 1β. Subunit 1β (also known as aryl hydrocarbon receptor nuclear translocator (ARNT)) is stable at any oxygen concentration, while subunit 1α is almost undetectable at normal oxygen concentrations because it is committed to proteosomal degradation [4]. Proteosomal degradation of HIF-1α is modulated by the prolyl hydroxylases 1, 2, and 3 (PHD1, PHD2, and PHD3, respectively), which hydroxylate proline residues 402 and 564 [5]. Hydroxylated subunit 1α is then recognized by the von Hippel–Lindau protein (pVHL), one of the components of the E3 ubiquitin ligase complex. This promotes its polyubiquitination and later destruction by 26s proteasome [5]. In contrast, under hypoxia HIF-1α is stabilized and translocates to the nucleus where it dimerizes with subunit 1β and activates target genes by binding to hypoxia-responsive elements (HREs) in the promoter region [4].

HIF-1 is responsible for the activation of over 60 genes that encode for growth factors, glucose transporters, transcription factors, erythropoietin, etc. In that manner, HIF-1 modulates oxygen consumption, cell survival, anaerobic metabolism, growth and development, and cell proliferation under hypoxic conditions [1,4].

During kidney development, HIF-1 appears to have important functions as well because it is present in many developing structures. For instance, HIF-1α protein is present in the nuclei of epithelial cells in ureteric buds in the medulla and cortex of human kidneys at gestational week 24. Moreover, it is found in epithelial cells of branching ampullae and S-shaped bodies in the nephrogenic zone. In rats and mice, HIF-1α is found predominantly in collecting ducts in the medulla, but rarely in the cortex [3,5]. *Vascular endothelial growth factor* (*VEGF*) is one main target gene of HIF-1 [3], and an essential molecule for normal renal development. For instance, the absence of the splicing isoforms $VEGF_{164}$ and $VEGF_{188}$ in mice has been associated with defective capillary angiogenesis, arteriogenesis, maturation of capillaries, development of glomerulosclerosis, and dilatation of proximal tubules and loops of Henle [6]. In addition, blocking VEGF in mice at postnatal day (PND) 0 caused a reduction on the number of nephrons and poor vascularity in the glomeruli [7]. Although the promoter region of *VEGF* has also consensus sites for Sp1/Sp3, AP-2, Egr-1, and STAT-3, HIF-1 seems to be a major determinant in the expression and secretion of VEGF during hypoxia [8]. These facts underline the importance of HIF-1 for proper kidney development.

Cadmium (Cd), a naturally occurring heavy metal, is a well-known nephrotoxicant that is capable of crossing the placenta to some extent and reaching the developing fetus [9,10]. Although a greater amount of Cd accumulates in the fetal liver, it can also reach fetal kidneys [10–12], and therefore, alter their development and cause damage.

A potential target for Cd toxicity during embryonic development is HIF-1. Several studies have shown that Cd may exert opposing effects on the mRNA expression and protein levels of HIF-1α, as well as on the ability of HIF-1 to bind to HREs in the promoter region of its target genes in several cell lines [13–17]. These differing effects could be attributed to the concentrations of Cd used, the time of exposure, and coexposure to a hypoxic stimulus. In animal models, those differences persist. For instance, in oysters, Cd exposure decreased *HIF-1α* and *PHD2* mRNA expression after postanoxic recovery [18], while under normoxia this metal ion only decreased *PHD2* mRNA expression [19]. In addition, in larval sheepshead minnow, cadmium reduced the hypoxia-induced mRNA expression of erythropoietin, a gene target of HIF-1 [20].

In spite of its nephrotoxic properties and its ability to cross the placenta, the effects of Cd exposure on embryonic kidneys are not well-investigated. Furthermore, the unavoidable exposure to this heavy metal through diet, pollution, and cigarette smoke [21] during the reproductive life stages emphasizes the importance of assessing the outcomes of the gestational exposure to Cd on this key regulator of embryonic development.

Therefore, the purpose of this study was to determine the effect of the intrauterine exposure of Wistar rats to Cd on HIF-1, its main regulator of protein stability, PHD2, and on its target gene, VEGF, in embryonic kidneys. The results show that Cd reduces HIF-1 DNA-binding ability, which was confirmed by a decrease in *VEGF* mRNA levels. However, VEGF protein levels were not altered, which suggests the activation of a compensation mechanism for appropriate kidney development.

2. Materials and Methods

2.1. Treatment of Animals and Tissue Collection

The Institutional Committee for the Care and Use of Laboratory Animals (Comité Interno para el Cuidado y uso de los Animales de Laboratorio, CICUAL) from CINVESTAV approved all animal procedures (protocol number: 041-13; approved date: 17 April 2013), and animal handling was performed in accordance with their guidelines. The samples used in this study were obtained from the animals employed in a previous study from our research group [12]. Briefly, pregnant Wistar rats were randomly divided in two groups: control (CT) and Cd, with 6 rats each. From GD 8 until GD 20, the rats from the CT group were exposed for 2 h/day to a mist of isotonic saline solution in a 10 L whole-body chamber connected to an Aeroneb nebulizer (inExpose, SCIREQ, Inc., Montreal, QC, Canada). Similarly, the rats from the Cd group were exposed by inhalation for 2 h/day to a mist of a solution of 1 mg CdCl$_2$/mL (Sigma-Aldrich Co., St. Louis, MO, USA). The nebulization was controlled with the flexiWare software v.6.1. (SCIREQ, Inc., Montreal, QC, Canada, 2012) to obtain a bias flow of 3 L/min, 10% nebulization rate (0.085 mL/min), and a nebulization cycle time of 1 s. The dose concentration of the aerosol (C$_{Dose}$) and the delivered dose (D$_{Del}$) achieved under these conditions were 17.43 mg Cd^{2+}/m^3 and 1.48 mg Cd^{2+}/kg/day, respectively. Cd exposure was applied through inhalation because it is an environmentally relevant route considering the permanent human exposure to this metal ion through air pollution and cigarette smoke [21]. Cadmium C$_{Dose}$, however, is high compared with the Cd concentration found in the environment [21]. This decision was made taking into consideration two aspects that could affect cadmium absorption: (1) the exposure period was rather short (13 days) in contrast to human exposure, and (2) aerosol particles' mass median aerodynamic diameter (MMAD) was larger than previous studies (2.5–3 μm) [22–24].

On GD 21, the dams were anesthetized with isoflurane (Sofloran Vet; PISA Farmacéutica, Hidalgo, Mexico) and the fetuses were obtained by caesarean section. The fetuses were kept in ice-cold isotonic saline solution until they were weighed and their kidneys obtained. Both kidneys of two different fetuses from each litter were homogenized independently in Trizol Reagent (Life Technologies, Carlsbad, CA, USA) the same day of extraction and homogenates were kept at −20 °C overnight until RNA extraction on the next morning.

The kidneys of the remaining fetuses were snap-frozen in liquid nitrogen and kept at −70 °C until protein extraction.

2.2. HIF-1 DNA-Binding Assay

A Procarta® Transcription Factor Plex Kit (Affymetrix, Inc., Santa Clara, CA, USA) was used to assess the activation of HIF-1. This assay is based on the Luminex® xMAP® Technology (Luminex Corp., Austin, TX, USA) and it is designed to measure DNA binding of transcription factors in nuclear extracts.

A pool of fetal kidneys from each litter was used to obtain nuclear extracts according to manufacturer's instructions. In addition, 96-well plates were prepared as indicated in the user's manual. Individual wells were loaded with 2 μg of nuclear extracts and each sample was run in duplicate. The plate was read in a Bio-Plex® System (Bio-Rad Laboratories, Inc., Hercules, CA, USA).

2.3. Quantitative Reverse Transcription Polymerase Chain Reaction (RT-qPCR)

As stated in Section 2.1, both kidneys from 2 fetuses from each litter were homogenized in Trizol Reagent (Life Technologies, Carlsbad, CA, USA). Homogenates were kept at −20 °C for the night, and on the next morning RNA extraction was performed according to manufacturer's indications. RNA integrity was assessed in 1.5% agarose gels electrophoresed at 90 V for 1 h and visualized with UV light. RNA samples were stored at −70 °C until use. cDNA was synthesized from 3 μg of RNA using the ImProm-II Reverse Transcription System Kit (Promega, Madison, WI, USA) according to the manufacturer's instructions and stored at −20 °C until its use. Purity and concentration of RNA and cDNA were evaluated using a NanoDrop 2000 (Thermo Fisher Scientific, Waltham, MA, USA).

Quantitative PCR was performed using the StepOne Real-Time PCR System (Applied Biosystems, Foster City, CA, USA) in 10 μL reactions containing: 5 μL of 2× mastermix with SYBR Green (Maxima SYBR Green/ROX qPCR Master Mix kit; Fermentas, Waltham, MA, USA), 0.1 μL of each primer solution (100 μM), 2.5 μg of cDNA and nuclease-free water to total 10 μL. Primer sequences used to assess *VEGF*, *PHD2*, HIF-1α, and *TATA-Binding Protein* (*TBP*) expression are as follows: *VEGF* 5′-TTACTGCTGTACCTCCAC-3′ (sense), 5′-ACAGGACGGCTTGAAGATA-3′ (anti-sense); *PHD2* 5′-CCATGGTCGCCTGTTACCC-3′ (sense), 5′-CGTACCTTGTGGCGTATGCAG-3′ (antisense); *HIF-1α* 5′-CCTACTATGTCGCTTTCTTGG-3′ (sense), 5′-TGTATGGGAGCATTAACTTCAC-3′ (antisense); *TBP* 5′-CACCGTGAATCTTGGCTGTAAAC-3′ (sense), 5′-CGCAGTTGTTCGTGGCTCTC-3′ (antisense). All primer sequences are published elsewhere [25–28]. Amplification was achieved according to the following cycling protocol: enzyme activation for 10 min at 95 °C, followed by 40 cycles of 15 s at 95 °C, 30 s at 60 °C, and 30 s at 72 °C. In order to assess the product specificity, a melting-curve analysis was performed.

Samples were normalized against TBP. The changes in expression relative to the CT group were calculated with the $2^{-\Delta\Delta CT}$ method. All samples were amplified in duplicate. The results are presented as mean fold changes of mRNA expression levels compared with controls, which were set to 1.0.

2.4. VEGF Enzyme-Linked Immunosorbent Assay (ELISA)

Protein levels of VEGF were quantified in total protein extracts obtained from a pool of fetal kidneys from each litter. A commercial ELISA kit was used (RAB0512, Sigma-Aldrich Co., St. Louis, MO, USA). Reconstitution and dilution of reagents, and plate preparation were performed according to the manufacturer's instructions. Twenty micrograms of total protein were used for each sample. All samples were run in duplicate. The color intensity of the samples was measured at 450 nm on a microplate reader (Infinite® F200, TECAN Group Ltd., Männedorf, Zürich, Switzerland). The concentration of VEGF is expressed as pg/mg protein.

2.5. Western Blot

Aliquots of the protein extracts used for the quantification of VEGF were used to assess the protein levels of PHD2 and HIF-1α.

Protein samples (30 μg/lane) were loaded onto 12% and 10% SDS-PAGE polyacrylamide gels for PHD2 and HIF-1α detection, respectively, and run at 0.04 A for 2 h. Proteins were transferred onto 0.45 μm pore-sized nitrocellulose membranes (Bio-Rad Laboratories, Hercules, CA, USA) for 2 h at 0.4 A for PHD2, and at 0.08 A for 16 h for HIF-1α. For PHD2, the membranes were blocked for 1 h at room temperature with 5% low-fat dry milk dissolved in 0.1% PBS-Tween 20, and for HIF-1α, membranes were blocked for 1.5 h. Membranes were incubated overnight at 4 °C with rabbit primary anti-PHD2 (4835, Cell Signaling Technology, Inc., Danvers, MA, USA; diluted 1:1000) or mouse primary anti-HIF-1α antibodies (NB-100-105, Novus Biologicals, Littleton, CO, USA; diluted 1:500) in 0.1% PBS-Tween 20. Blots were washed and incubated for 1 h at room temperature with goat anti-rabbit secondary antibody (sc-2004, Santa Cruz Biotechnology, Inc., Dallas, TX, USA; diluted 1:10,000 in 0.1% PBS-tween 20) or 1.5 h with the goat antimouse secondary antibody (sc-2005, Santa Cruz Biotechnology, Inc., Dallas, TX, USA; diluted 1:10,000 in 0.1% PBS-Tween 20). After washing, proteins were detected by chemiluminescence (LuminataTM Forte, Millipore Corp., Burlington, MA, USA) using the LI-COR C-DiGit scanner (LI-COR, Inc., Lincoln, NE, USA).

Densitometric analysis was performed using the Image Studio Lite Software v.5.0.21 (LI-COR, Inc., Lincoln, NE, USA) using β-actin as a loading control.

2.6. Statisticals

Statistical analyses were performed with the GraphPad Prism software version 7.0 for Windows (GraphPad software, La Jolla, CA, USA). Means ± standard error of the mean (SEM) are shown. Student's *t*-test (unpaired, two-tailed) or Mann–Whitney U-test (unpaired, two-tailed) were performed for parametric and nonparametric data, respectively. $p \leq 0.05$ was considered statistically significant.

3. Results

3.1. Effect of Intrauterine Cadmium Exposure on DNA-Binding Ability of HIF-1 in Fetal Kidneys

Because HIF-1 is a transcription factor, its ability to bind specific sequences in the DNA is essential for gene regulation. In this study, the activity of HIF-1 was evaluated with a multiplex assay in which the samples showed higher fluorescence intensity when HIF-1 binding increased.

Cadmium exposure during gestation significantly reduced the ability of HIF-1 to bind DNA in fetal kidneys by 44 ± 11% (Figure 1).

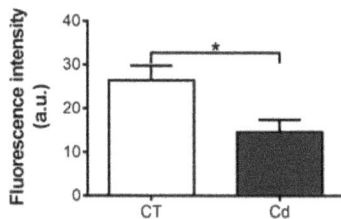

Figure 1. Hypoxia-inducible factor 1 (HIF-1) DNA-binding ability in nuclear extracts of kidneys of control and cadmium-exposed fetuses. Two micrograms of nuclear extracts were used per sample. Each sample was run in duplicate. The bars represent means ± SEM, *n* = 5. * *p* = 0.0307, Student's *t*-test.

3.2. Effect of Cadmium on the mRNA Expression of VEGF-A, PHD2, and HIF-1α in Fetal Kidneys

The mRNA expression of *VEGF* was assessed since it is an important target gene of HIF-1, which has been associated with normal kidney development. Cadmium exposure significantly decreased the expression of *VEGF* by 44 ± 7% in fetal kidneys compared with the CT group (Figure 2a). However, Cd exposure did not alter the expression of *HIF-1α* or of *PHD2*, which promotes HIF-1α degradation (Figure 2b,c).

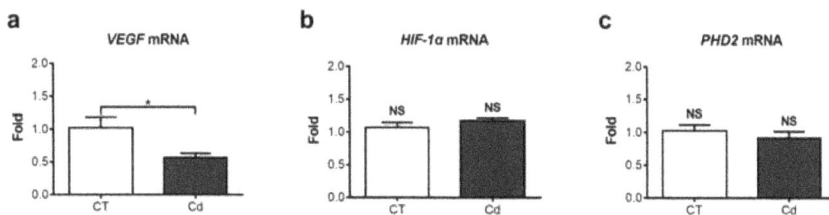

Figure 2. mRNA expression of (**a**) vascular endothelial growth factor (*VEGF*) (* *p* = 0.0266, Mann–Whitney U-test), (**b**) *HIF-1α* (*p* = 0.1379, Mann–Whitney U-test), and (**c**) *PHD2* (*p* = 0.4083, Student's *t*-test) in kidneys of fetuses from control (CT) and cadmium (Cd) groups. Data were normalized with *TATA-Binding Protein (TBP)* expression. Each sample was run in duplicate. The fold changes compared to the CT group are plotted. Means ± SEM are shown, *n* = 9. NS, not statistically significant.

3.3. Cadmium-Induced Changes of VEGF, PHD2, and HIF-1α Protein in Fetal Kidneys

To complement gene expression, protein levels of VEGF, PHD2, and HIF-1α were also assessed. In contrast to expectation, VEGF protein levels in the kidneys of Cd-exposed fetuses (759.2 ± 184.7 pg/mg total protein) did not differ from those of the control group (548.8 ± 98.9 pg/mg total protein) (Figure 3a). Similarly, the densitometric analysis showed no difference in the relative levels of PHD2 and HIF-1α of the CT and Cd groups (Figure 3b,c), which was in accordance with the findings of their gene expression.

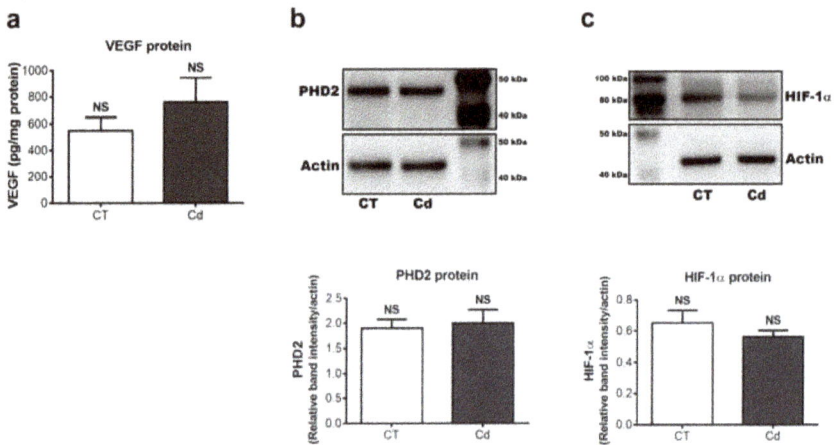

Figure 3. Protein levels of (**a**) VEGF (*p* = 0.3389, Student's *t*-test), (**b**) PHD2 (*p* = 0.7702, Student's *t*-test), and (**c**) HIF-1α (*p* = 0.3476, Student's *t*-test) in renal tissue from control and Cd-exposed fetuses. (**b**,**c**) show representative blots of one of six samples from each group, and densitometry. Band intensities were normalized to actin level. VEGF concentration was normalized to protein content. Each sample was run in duplicate. Means ± SEM are shown, *n* = 6. NS, not statistically significant.

4. Discussion

The long-term exposure, even at low doses, of pregnant women to heavy metals capable of accumulating in the body can generate irreversible outcomes in fetal growth and development [29]. Cd intoxication is a common danger for all organisms because of its continuing presence in polluted air and tobacco smoke that, when inhaled by pregnant women, poses a serious threat to the woman and, particularly, the developing fetus due to lack of (or minimal presence of) mechanisms of protection [29].

Because of this, it is important to characterize the effects of gestational exposure of Cd metal ion and find its possible targets. As stated in the introduction section, one of them is HIF-1, one of the main transcription factors that control the expression of several genes that are necessary for cell survival and proliferation, and glucose metabolism [4] under hypoxic conditions, such as during embryonic development.

The expression of its subunit 1α, as well as its activity, can be modified by the exposure to several metal ions, like cobalt, nickel [14,30,31], and Cd. The aforementioned metal ions increase the mRNA and/or protein levels of HIF-1α, as well as the ability of HIF-1 to bind to its HREs. For Cd, however, conflicting results have been obtained. Several studies in cell lines, as well as animal models, have shown that Cd increases HIF-1α protein levels and HIF-1 activity, as reflected by a rise in VEGF transcription [16,32]. Nevertheless, in HEK293 and Hep3B cells under a hypoxic stimulus, Cd decreased the ability of HIF-1 to bind DNA [13,15,17] because of increased proteasomal degradation of subunit 1α [13].

In the light of these observations, the goal of the study was to assess the effects of Cd exposure on HIF-1 in kidney tissue during a highly vulnerable stage, namely, gestational development.

The effectiveness of the parameters of exposure (described in Section 2.1) was evaluated by measuring the content of this metal ion in dam and fetal organs such as lungs, liver, kidneys, placentae, and fetal kidneys. The cadmium burden on those organs was significantly greater than in the control group [12], demonstrating an effective absorption and distribution of Cd in the dam, and, more importantly, showing that this transition metal ion crosses the placenta and reaches developing kidneys. Moreover, dam organs did not show any changes in relative weight, and mothers did not present signs of overt toxicity [12], suggesting that all changes observed in fetal kidneys were direct effects of Cd.

This study presents evidence that gestational exposure to Cd decreases HIF-1 DNA-binding ability in embryonic kidneys without altering the mRNA or protein levels of its subunit 1α, which is in accordance with the lack of change in PHD2 mRNA and protein levels.

So far, the mechanism by which Cd could be exerting its inhibitory effect on HIF-1 activity without modifying HIF-1α protein levels has not been elucidated. However, Kubis et al. [33] reported a cytoplasmic accumulation of the subunit 1α in a skeletal muscle primary culture from New Zealand rabbits exposed to Cd and subjected to hypo-(3% O_2) and hyperoxic (42% O_2) conditions. This caused a decrease of the translocation of HIF-1α to the nucleus and lower mRNA levels of *glyceraldehyde-3-phosphate dehydrogenase* (*GAPDH*), one of HIF-1's target genes. The authors attributed the lower nuclear import and, therefore, DNA-binding ability to a higher association of subunit 1α to heat shock protein 90 (Hsp90). Hsp90 is a chaperone that prevents the aggregation of un- or misfolded proteins produced under stress situations [34]. Additionally, it takes part in von Hippel–Lindau-independent regulation of HIF-1. Under normoxic conditions, Hsp90 binds to the basic helix-loop-helix-Per-ARNT-Sim (bHLHL-PAS) domain found in subunit 1α, which stabilizes the protein and prevents it from being degraded, but keeps the subunit in an inactive state. Under hypoxic conditions, the binding loses its strength, allowing the nuclear translocation of the subunit and further binding to its HREs [34–36].

In this study, we assessed neither Hsp90 levels nor its binding to subunit 1α in embryonic kidneys, but previous reports indicate that basal levels of this protein in newborn kidneys from both humans and rats (PND 1) is higher than in adult kidneys [37,38]. In fetal kidneys, Hsp90 is expressed in the parietal epithelium of Bowman's capsule, podocytes, blastema, S-shaped bodies, proximal convoluted and straight tubules, as well as collecting ducts [37,38]. Furthermore, several studies associated Cd exposure with increased Hsp90 levels. For instance, Leghorn chick embryos exposed *in ovo* to Cd for 24 h had higher protein levels of Hsp24, Hsp70, and Hsp90 than those in the control group [39]. Similar increases were found in renal tissue from rat [40], duck [41], carp [42], and the proximal tubule cell line, NRK-52E exposed to Cd [43]. More importantly, a study showed that increasing doses of Cd (0.5–4 mg Cd^{2+}/kg, i.p.) resulted in a dose-dependent increase in the level of association between Hsp90 and one of its substrates, the glucocorticoid receptor, in liver cytosolic extracts [44]. Putting these findings together, it is plausible that Cd increases the degree of association between subunit 1α and Hsp90 by a process (still to be identified) without altering the protein levels of 1α subunit, thus leading to a cytoplasmic accumulation of the subunit and lower DNA-binding ability of HIF-1. This hypothesis needs to be tested in our model of exposure by determining HIF-1α nuclear and cytoplasmic levels, as well as its association to HSP90.

Another possible explanation relies on the fact that HIF-1 requires copper to bind its HREs because it promotes complexation of HIF-1α with its cofactor p300 through inhibition of the factor inhibiting HIF-1 (FIH-1) [45]. It has been shown that Cd exposure decreases the maternal transfer of micronutrients such as iron, zinc, and copper, thus altering their burden in several organs [46,47]. This raises the possibility that HIF-1 activity may be decreased by a lower copper content in fetal kidneys, as previously reported [48]. This speculation requires, of course, further experimental confirmation since contrasting effects have been found on this matter [46,48,49], probably due to dose, route, length, and period of exposure.

Consistent with a lower HIF-1 DNA-binding ability, *VEGF* mRNA levels were significantly reduced by Cd exposure, which confirms a reduced activity of this transcription factor. This Cd-induced effect was previously reported by Gheorghescu et al. [50]. The authors showed that the exposure of chick embryos from the Ross strain to a 50 μM Cd acetate solution decreased *VEGF-A* mRNA expression in extraembryonic membranes 1 h post-treatment [50]. In contrast to the mRNA levels, VEGF-A protein expression was not modified by Cd exposure. This difference could be due to pleiotropic regulation of this growth factor. At the translational level, VEGF is regulated by internal ribosome entry sites, upstream open reading frames, alternative initiation codons, micro-RNAs, riboswitches, and RNA G-quadruplex structures [51]. This is understandable considering the importance of this molecule for angiogenesis and vasculogenesis and, therefore, fetal development. Thus, it is possible that the rate of protein translation was increased or post-translational modifications of VEGF affected by any of those mechanisms to compensate the lowered mRNA levels induced by Cd. Nevertheless, further corroboration is needed.

It is noteworthy that the present study serves as a first approach to try to elucidate molecular targets of cadmium in developing organisms; hence, it is important to do further assessments that will help to understand more clearly the detrimental effects (not only renal) of this transition metal after gestational exposure. This should include the evaluation of the relative expression of other HIF-1 target genes, as well as their protein levels and the degree of their functionality, if possible. The expression of miRNAs should also be evaluated, since they participate in the post-transcriptional regulation of gene expression and might have an important role during renal (and embryonic overall) development. In addition, an alternative approach could include the use of cellular models from embryonic origin to confirm these findings and evaluate a possible mechanism by which cadmium is impairing HIF-1 DNA-binding, as well as other molecular targets.

5. Conclusions

The results of this study show that Cd in utero exposure impairs HIF-1 DNA-binding activity in developing kidneys, by a mechanism that is independent of HIF-1α protein levels and needs to be identified. This observation is in accordance with previous reports. In addition, reduced transcriptional activity of HIF-1 was confirmed by lower *VEGF* mRNA levels although protein levels remained unchanged. This finding suggests the existence of alternative compensatory mechanisms to maintain adequate protein levels of this key molecule and thus ensure proper fetal development. Nevertheless, it is important to study further possible outcomes of decreased HIF-1 activity, as well as other mechanisms of Cd toxicity targeting HIF-1, because embryonic development is highly vulnerable, and any alteration during this stage can lead to developmental defects later in life.

Author Contributions: O.B. and T.J.-E. conceived and designed the experiments; M.C.-G. and T.J.-E. performed the experiments; all the authors analyzed the data and wrote the paper.

Funding: This study was supported by Secretaría de Ciencia, Tecnología e Innovación (SECITI, grant PICSA12-086), and it is the result of collaborative work between research groups in Mexico and Germany, which is supported by a joint grant Conacyt-BMBF (Conacyt 267755; BMBF 01DN16039). Tania Jacobo-Estrada was recipient of a fellowship from CONACyT (grant 326697).

Acknowledgments: We wish to thank Biol. Gabriel Vargas Corona for his technical assistance and Aurora Espejel-Nuñez (Departamento de Inmunobioquímica, Instituto Nacional de Perinatología, Ciudad de México, Mexico) for lending us the Bio-Plex System and for her advice for reading the plate.

Conflicts of Interest: The authors declare no conflict of interest. The funding sponsors had no role in the design of the study; in the collection, analyses, or interpretation of data; in the writing of the manuscript; and in the decision to publish the results.

References

1. Fajersztajn, L.; Veras, M.M. Hypoxia: From placental development to fetal programming. *Birth Defects Res.* **2017**, *109*, 1377–1385. [CrossRef] [PubMed]
2. Beaudoin, A.R. Embriology and teratology. In *The Laboratory Rat*; Baker, H.J., Lindsey, J.R., Weisbroth, S.H., Eds.; Academic Press: New York, NY, USA, 1979; Volume 2, pp. 75–101.
3. Bernhardt, W.M.; Schmitt, R.; Rosenberger, C.; Munchenhagen, P.M.; Grone, H.J.; Frei, U.; Warnecke, C.; Bachmann, S.; Wiesener, M.S.; Willam, C.; et al. Expression of hypoxia-inducible transcription factors in developing human and rat kidneys. *Kidney Int.* **2006**, *69*, 114–122. [CrossRef] [PubMed]
4. Dery, M.A.; Michaud, M.D.; Richard, D.E. Hypoxia-inducible factor 1: Regulation by hypoxic and non-hypoxic activators. *Int. J. Biochem. Cell Biol.* **2005**, *37*, 535–540. [CrossRef] [PubMed]
5. Gunaratnam, L.; Bonventre, J.V. HIF in kidney disease and development. *J. Am. Soc. Nephrol.* **2009**, *20*, 1877–1887. [CrossRef] [PubMed]
6. Mattot, V.; Moons, L.; Lupu, F.; Chernavvsky, D.; Gomez, R.A.; Collen, D.; Carmeliet, P. Loss of the VEGF(164) and VEGF(188) isoforms impairs postnatal glomerular angiogenesis and renal arteriogenesis in mice. *J. Am. Soc. Nephrol.* **2002**, *13*, 1548–1560. [CrossRef] [PubMed]
7. Kitamoto, Y.; Tokunaga, H.; Tomita, K. Vascular endothelial growth factor is an essential molecule for mouse kidney development: Glomerulogenesis and nephrogenesis. *J. Clin. Investig.* **1997**, *99*, 2351–2357. [CrossRef] [PubMed]
8. Pagès, G.; Pouyssegur, J. Transcriptional regulation of the vascular endothelial growth factor gene— A concert of activating factors. *Cardiovasc. Res.* **2005**, *65*, 564–573. [CrossRef] [PubMed]
9. Nakamura, Y.; Ohba, K.; Ohta, H. Participation of metal transporters in cadmium transport from mother rat to fetus. *J. Toxicol. Sci.* **2012**, *37*, 1035–1044. [CrossRef] [PubMed]
10. Nakamura, Y.; Ohba, K.; Suzuki, K.; Ohta, H. Health effects of low-level cadmium intake and the role of metallothionein on cadmium transport from mother rats to fetus. *J. Toxicol. Sci.* **2012**, *37*, 149–156. [CrossRef] [PubMed]
11. Jacquillet, G.; Barbier, O.; Rubera, I.; Tauc, M.; Borderie, A.; Namorado, M.C.; Martin, D.; Sierra, G.; Reyes, J.L.; Poujeol, P.; et al. Cadmium causes delayed effects on renal function in the offspring of cadmium-contaminated pregnant female rats. *Am. J. Physiol. Ren. Physiol.* **2007**, *293*, F1450–F1460. [CrossRef] [PubMed]
12. Jacobo-Estrada, T.; Cardenas-Gonzalez, M.; Santoyo-Sanchez, M.; Parada-Cruz, B.; Uria-Galicia, E.; Arreola-Mendoza, L.; Barbier, O. Evaluation of kidney injury biomarkers in rat amniotic fluid after gestational exposure to cadmium. *J. Appl. Toxicol.* **2016**, *36*, 1183–1193. [CrossRef] [PubMed]
13. Chun, Y.S.; Choi, E.; Kim, G.T.; Choi, H.; Kim, C.H.; Lee, M.J.; Kim, M.S.; Park, J.W. Cadmium blocks hypoxia-inducible factor (HIF)-1-mediated response to hypoxia by stimulating the proteasome-dependent degradation of HIF-1alpha. *Eur. J. Biochem.* **2000**, *267*, 4198–4204. [CrossRef] [PubMed]
14. Gao, S.; Zhou, J.; Zhao, Y.; Toselli, P.; Li, W. Hypoxia-response element (hre)-directed transcriptional regulation of the rat lysyl oxidase gene in response to cobalt and cadmium. *Toxicol. Sci.* **2013**, *132*, 379–389. [CrossRef] [PubMed]
15. Horiguchi, H.; Kayama, F.; Oguma, E.; Willmore, W.G.; Hradecky, P.; Bunn, H.F. Cadmium and platinum suppression of erythropoietin production in cell culture: Clinical implications. *Blood* **2000**, *96*, 3743–3747. [PubMed]
16. Jing, Y.; Liu, L.Z.; Jiang, Y.; Zhu, Y.; Guo, N.L.; Barnett, J.; Rojanasakul, Y.; Agani, F.; Jiang, B.H. Cadmium increases HIF-1 and VEGF expression through ROS, ERK, and AKT signaling pathways and induces malignant transformation of human bronchial epithelial cells. *Toxicol. Sci.* **2012**, *125*, 10–19. [CrossRef] [PubMed]
17. Obara, N.; Imagawa, S.; Nakano, Y.; Suzuki, N.; Yamamoto, M.; Nagasawa, T. Suppression of erythropoietin gene expression by cadmium depends on inhibition of HIF-1, not stimulation of GATA-2. *Arch. Toxicol.* **2003**, *77*, 267–273. [CrossRef] [PubMed]
18. Ivanina, A.V.; Sokolov, E.P.; Sokolova, I.M. Effects of cadmium on anaerobic energy metabolism and mRNA expression during air exposure and recovery of an intertidal mollusk Crassostrea virginica. *Aquat. Toxicol.* **2010**, *99*, 330–342. [CrossRef] [PubMed]

19. Piontkivska, H.; Chung, J.S.; Ivanina, A.V.; Sokolov, E.P.; Techa, S.; Sokolova, I.M. Molecular characterization and mRNA expression of two key enzymes of hypoxia-sensing pathways in eastern oysters Crassostrea virginica (Gmelin): Hypoxia-inducible factor alpha (HIF-alpha) and HIF-prolyl hydroxylase (PHD). *Comp. Biochem. Physiol. Part D Genomics Proteomics* **2011**, *6*, 103–114. [CrossRef] [PubMed]

20. Dangre, A.J.; Manning, S.; Brouwer, M. Effects of cadmium on hypoxia-induced expression of hemoglobin and erythropoietin in larval sheepshead minnow, cyprinodon variegatus. *Aquat. Toxicol.* **2010**, *99*, 168–175. [CrossRef] [PubMed]

21. Agency for Toxic Substances and Disease Registry. *Toxicological Profile for Cadmium*; Department of Health and Human Services, Public Health Service: Atlanta, GA, USA, 2012.

22. Prigge, E. Early signs of oral and inhalative cadmium uptake in rats. *Arch. Toxicol.* **1978**, *40*, 231–247. [CrossRef] [PubMed]

23. Baranski, B. Behavioral alterations in offspring of female rats repeatedly exposed to cadmium oxide by inhalation. *Toxicol. Lett.* **1984**, *22*, 53–61. [CrossRef]

24. Trottier, B.; Athot, J.; Ricard, A.C.; Lafond, J. Maternal-fetal distribution of cadmium in the guinea pig following a low dose inhalation exposure. *Toxicol. Lett.* **2002**, *129*, 189–197. [CrossRef]

25. Cabiati, M.; Raucci, S.; Caselli, C.; Guzzardi, M.A.; D'Amico, A.; Prescimone, T.; Giannessi, D.; Del Ry, S. Tissue-specific selection of stable reference genes for real-time pcr normalization in an obese rat model. *J. Mol. Endocrinol.* **2012**, *48*, 251–260. [CrossRef] [PubMed]

26. Chen, Y.; Fu, L.; Han, Y.; Teng, Y.; Sun, J.; Xie, R.; Cao, J. Testosterone replacement therapy promotes angiogenesis after acute myocardial infarction by enhancing expression of cytokines HIF-1a, SDF-1a and VEGF. *Eur. J. Pharmacol.* **2012**, *684*, 116–124. [CrossRef] [PubMed]

27. Chen, Y.R.; Dai, A.G.; Hu, R.C.; Jiang, Y.L. Differential and reciprocal regulation between hypoxia-inducible factor-alpha subunits and their prolyl hydroxylases in pulmonary arteries of rat with hypoxia-induced hypertension. *Acta Biochim. Biophys. Sin.* **2006**, *38*, 423–434. [CrossRef] [PubMed]

28. Katavetin, P.; Miyata, T.; Inagi, R.; Tanaka, T.; Sassa, R.; Ingelfinger, J.R.; Fujita, T.; Nangaku, M. High glucose blunts vascular endothelial growth factor response to hypoxia via the oxidative stress-regulated hypoxia-inducible factor/hypoxia-responsible element pathway. *J. Am. Soc. Nephrol.* **2006**, *17*, 1405–1413. [CrossRef] [PubMed]

29. Semczuk, M.; Semczuk-Sikora, A. New data on toxic metal intoxication (Cd, Pb, and Hg in particular) and mg status during pregnancy. *Med. Sci. Monit.* **2001**, *7*, 332–340. [PubMed]

30. Salnikow, K.; Su, W.; Blagosklonny, M.V.; Costa, M. Carcinogenic metals induce hypoxia-inducible factor-stimulated transcription by reactive oxygen species-independent mechanism. *Cancer Res.* **2000**, *60*, 3375–3378. [PubMed]

31. Yao, Y.X.; Lu, Y.H.; Chen, W.C.; Jiang, Y.P.; Cheng, T.; Ma, Y.P.; Lu, L.; Dai, W. Cobalt and nickel stabilize stem cell transcription factor oct4 through modulating its sumoylation and ubiquitination. *PLoS ONE* **2014**, *9*. [CrossRef] [PubMed]

32. Liu, F.; Wang, B.; Li, L.; Dong, F.; Chen, X.; Li, Y.; Dong, X.; Wada, Y.; Kapron, C.M.; Liu, J. Low-dose cadmium upregulates vegf expression in lung adenocarcinoma cells. *Int. J. Environ. Res. Public Health* **2015**, *12*, 10508–10521. [CrossRef] [PubMed]

33. Kubis, H.P.; Hanke, N.; Scheibe, R.J.; Gros, G. Accumulation and nuclear import of HIF1 alpha during high and low oxygen concentration in skeletal muscle cells in primary culture. *Biochim. Biophys. Acta* **2005**, *1745*, 187–195. [CrossRef] [PubMed]

34. Minet, E.; Mottet, D.; Michel, G.; Roland, I.; Raes, M.; Remacle, J.; Michiels, C. Hypoxia-induced activation of HIF-1: Role of HIF-1alpha-hsp90 interaction. *FEBS Lett.* **1999**, *460*, 251–256. [CrossRef]

35. Isaacs, J.S.; Jung, Y.J.; Mimnaugh, E.G.; Martinez, A.; Cuttitta, F.; Neckers, L.M. Hsp90 regulates a von hippel lindau-independent hypoxia-inducible factor-1 alpha-degradative pathway. *J. Biol. Chem.* **2002**, *277*, 29936–29944. [CrossRef] [PubMed]

36. Zhang, D.; Li, J.; Costa, M.; Gao, J.; Huang, C. JNK1 mediates degradation HIF-1alpha by a VHL-independent mechanism that involves the chaperones Hsp90/Hsp70. *Cancer Res.* **2010**, *70*, 813–823. [CrossRef] [PubMed]

37. Somji, S.; Ann Sens, M.; Garrett, S.H.; Gurel, V.; Todd, J.H.; Sens, D.A. Expression of Hsp90 in the human kidney and in proximal tubule cells exposed to heat, sodium arsenite and cadmium chloride. *Toxicol. Lett.* **2002**, *133*, 241–254. [CrossRef]

38. D'Souza, S.M.; Brown, I.R. Constitutive expression of heat shock proteins Hsp90, Hsc70, Hsp70 and Hsp60 in neural and non-neural tissues of the rat during postnatal development. *Cell Stress Chaperones* **1998**, *3*, 188–199. [CrossRef]
39. Papaconstantinou, A.D.; Brown, K.M.; Noren, B.T.; McAlister, T.; Fisher, B.R.; Goering, P.L. Mercury, cadmium, and arsenite enhance heat shock protein synthesis in chick embryos prior to embryotoxicity. *Birth Defects Res. B Dev. Reprod. Toxicol.* **2003**, *68*, 456–464. [CrossRef] [PubMed]
40. Goering, P.L.; Kish, C.L.; Fisher, B.R. Stress protein synthesis induced by cadmium-cysteine in rat kidney. *Toxicology* **1993**, *85*, 25–39. [CrossRef]
41. Xia, B.; Cao, H.; Luo, J.; Liu, P.; Guo, X.; Hu, G.; Zhang, C. The co-induced effects of molybdenum and cadmium on antioxidants and heat shock proteins in duck kidneys. *Biol. Trace Elem. Res.* **2015**, *168*, 261–268. [CrossRef] [PubMed]
42. Hermesz, E.; Abraham, M.; Nemcsok, J. Identification of two Hsp90 genes in carp. *Comp. Biochem. Physiol. C Toxicol. Pharmacol.* **2001**, *129*, 397–407. [CrossRef]
43. Madden, E.F.; Akkerman, M.; Fowler, B.A. A comparison of 60, 70, and 90 kDa stress protein expression in normal rat NRK-52 and human HK-2 kidney cell lines following in vitro exposure to arsenite and cadmium alone or in combination. *J. Biochem. Mol. Toxicol.* **2002**, *16*, 24–32. [CrossRef] [PubMed]
44. Dundjerski, J.; Kovac, T.; Pavkovic, N.; Cvoro, A.; Matic, G. Glucocorticoid receptor-Hsp90 interaction in the liver cytosol of cadmium-intoxicated rats. *Cell Biol. Toxicol.* **2000**, *16*, 375–383. [CrossRef] [PubMed]
45. Feng, W.; Ye, F.; Xue, W.; Zhou, Z.; Kang, Y.J. Copper regulation of hypoxia-inducible factor-1 activity. *Mol. Pharmacol.* **2009**, *75*, 174–182. [CrossRef] [PubMed]
46. Kuriwaki, J.; Nishijo, M.; Honda, R.; Tawara, K.; Nakagawa, H.; Hori, E.; Nishijo, H. Effects of cadmium exposure during pregnancy on trace elements in fetal rat liver and kidney. *Toxicol. Lett.* **2005**, *156*, 369–376. [CrossRef] [PubMed]
47. Petering, H.G.; Choudhury, H.; Stemmer, K.L. Some effects of oral ingestion of cadmium on zinc, copper, and iron metabolism. *Environ. Health Perspect.* **1979**, *28*, 97–106. [CrossRef] [PubMed]
48. Sowa, B.; Steibert, E. Effect of oral cadmium administration to female rats during pregnancy on zinc, copper, and iron content in placenta, foetal liver, kidney, intestine, and brain. *Arch. Toxicol.* **1985**, *56*, 256–262. [CrossRef] [PubMed]
49. Baranski, B. Effect of cadmium on prenatal development and on tissue cadmium, copper, and zinc concentrations in rats. *Environ. Res.* **1987**, *42*, 54–62. [CrossRef]
50. Gheorghescu, A.K.; Tywoniuk, B.; Duess, J.; Buchete, N.V.; Thompson, J. Exposure of chick embryos to cadmium changes the extra-embryonic vascular branching pattern and alters expression of VEGF-A and VEGF-R2. *Toxicol. Appl. Pharmacol.* **2015**, *289*, 79–88. [CrossRef] [PubMed]
51. Arcondeguy, T.; Lacazette, E.; Millevoi, S.; Prats, H.; Touriol, C. VEGF-A mrna processing, stability and translation: A paradigm for intricate regulation of gene expression at the post-transcriptional level. *Nucleic Acids Res.* **2013**, *41*, 7997–8010. [CrossRef] [PubMed]

Article

Cadmium Nephrotoxicity Is Associated with Altered MicroRNA Expression in the Rat Renal Cortex

Michael J. Fay [1,2,]* , **Lauren A. C. Alt [1]**, **Dominika Ryba [1]**, **Ribhi Salamah [1]**, **Ryan Peach [1]**,
Alexander Papaeliou [1], **Sabina Zawadzka [1]**, **Andrew Weiss [1]**, **Nil Patel [1]**, **Asad Rahman [1]**,
Zyaria Stubbs-Russell [1], **Peter C. Lamar [2]**, **Joshua R. Edwards [2]** and **Walter C. Prozialeck [2]**

[1]   Department of Biomedical Sciences, Midwestern University, 555 31st Street, Downers Grove, IL 60515, USA;
      lalt@midwestern.edu (L.A.C.A.); dryba24@midwestern.edu (D.R.); rsalamah66@midwestern.edu (R.S.);
      rpeach38@midwestern.edu (R.P.); apapaeliou97@midwestern.edu (A.P.);
      szawadzka34@midwestern.edu (S.Z.); aweiss16@midwestern.edu (A.W.); npatel16@midwestern.edu (N.P.);
      arahman13@midwestern.edu (A.R.); zstubbs-russell11@midwestern.edu (Z.S.-R.)
[2]   Department of Pharmacology, Midwestern University, 555 31st Street, Downers Grove, IL 60515, USA;
      plamar@midwestern.edu (P.C.L.); jedwar@midwestern.edu (J.R.E.); wprozi@midwestern.edu (W.C.P.)
*   Correspondence: mfayxx@midwestern.edu; Tel.: +1-630-515-6382

Received: 28 February 2018; Accepted: 9 March 2018; Published: 15 March 2018

Abstract: Cadmium (Cd) is a nephrotoxic environmental pollutant that causes a generalized dysfunction of the proximal tubule characterized by polyuria and proteinuria. Even though the effects of Cd on the kidney have been well-characterized, the molecular mechanisms underlying these effects have not been fully elucidated. MicroRNAs (miRNAs) are small non-coding RNAs that regulate cellular and physiologic function by modulating gene expression at the post-transcriptional level. The goal of the present study was to determine if Cd affects renal cortex miRNA expression in a well-established animal model of Cd-induced kidney injury. Male Sprague-Dawley rats were treated with subcutaneous injections of either isotonic saline or $CdCl_2$ (0.6 mg/kg) 5 days a week for 12 weeks. The 12-week Cd-treatment protocol resulted in kidney injury as determined by the development of polyuria and proteinuria, and a significant increase in the urinary biomarkers Kim-1, β_2 microglobulin and cystatin C. Total RNA was isolated from the renal cortex of the saline control and Cd treated animals, and differentially expressed miRNAs were identified using µParaflo™ microRNA microarray analysis. The microarray results demonstrated that the expression of 44 miRNAs were significantly increased and 54 miRNAs were significantly decreased in the Cd treatment group versus the saline control (t-test, $p \leq 0.05$, $N = 6$ per group). miR-21-5p, miR-34a-5p, miR-146b-5p, miR-149-3p, miR-224-5p, miR-451-5p, miR-1949, miR-3084a-3p, and miR-3084c-3p demonstrated more abundant expression and a significant two-fold or greater increased expression in the Cd-treatment group versus the saline control group. miR-193b-3p, miR-455-3p, and miR-342-3p demonstrated more abundant expression and a significant two-fold or greater decreased expression in the Cd-treatment group versus the saline control group. Real-time PCR validation demonstrated (1) a significant (t-test, $p \leq 0.05$, $N = 6$ per group) increase in expression in the Cd-treated group for miR-21-5p (2.7-fold), miR-34a-5p (10.8-fold), miR-146b-5p (2-fold), miR-224-5p (10.2-fold), miR-3084a-3p (2.4-fold), and miR-3084c-3p (3.3-fold) and (2) a significant (t-test, $p \leq 0.05$, $N = 6$ per group) 52% decrease in miR-455-3p expression in the Cd-treatment group. These findings demonstrate that Cd significantly alters the miRNA expression profile in the renal cortex and raises the possibility that dysregulated miRNA expression may play a role in the pathophysiology of Cd-induced kidney injury. In addition, these findings raise the possibility that Cd-dysregulated miRNAs might be used as urinary biomarkers of Cd exposure or Cd-induced kidney injury.

Keywords: cadmium; microRNAs; nephrotoxicity; biomarkers

1. Introduction

The nephrotoxic heavy metal cadmium (Cd) is a Group 1 carcinogen and currently ranked 7th on the 2017 Agency for Toxic Substances and Disease Registry (ATSDR) and EPA list of hazardous substances [1]. Industrial activities have resulted in increases in the concentrations of Cd in the environment. Human exposure can occur by inhalation in the workplace, the ingestion of contaminated food and water, and smoking tobacco [2]. Circulating Cd that is bound to low-molecular-weight proteins or thiol compounds is filtered at the glomerulus and taken up by proximal tubule epithelial cells, and chronic low-level human exposure to Cd results in proximal tubule accumulation [3]. When a critical Cd threshold of 150–200 µg/g wet weight (equivalent to 450–600 µg/g dry weight) is reached, toxic injury can occur, which is manifested by a generalized reabsorptive dysfunction resulting in polyuria and low-molecular-weight proteinuria [4,5].

While the toxic effects of Cd on the proximal tubule are well documented, the molecular mechanisms associated with Cd-induced kidney injury have not been fully elucidated. Prior to causing cell death, Cd has been shown to induce oxidative stress, promote cytoskeletal reorganization, disrupt cadherin-dependent cell–cell adhesion, decrease transepithelial electrical resistance, activate various cellular signaling pathways, and induce endoplasmic reticulum stress and autophagy in proximal tubule epithelial cells [6,7]. Even though Cd can affect a wide variety of cellular processes, it is not well established how novel regulators such as microRNAs (miRNAs) might be involved in Cd-induced proximal tubule epithelial cellular injury.

MicroRNAs are evolutionarily conserved small (20–25 nt) non-protein coding RNAs that inhibit gene expression at the post-transcriptional level by blocking mRNA translation or promoting mRNA degradation within a cell [8,9]. These non-protein coding miRNAs can have a major impact on cellular function, as a single miRNA can interact with hundreds of different protein-coding mRNAs, and a single protein-coding mRNA can be affected by multiple miRNAs [10,11]. MicroRNAs are known to be involved in kidney development, kidney homeostasis, and the pathophysiology of kidney disease [12–16]. There is also evidence that dysregulated miRNA expression is associated with kidney injury in both rodent and human studies, and that miRNAs have the potential to serve as urinary biomarkers of kidney injury [14,16–20]. However, there is a lack of information concerning the role of miRNAs in Cd-induced nephrotoxicity. The purpose of this research study was to use a well-established sub-chronic animal model to determine if Cd-induced nephrotoxicity is associated with dysregulated miRNA expression in the renal cortex [21].

2. Material and Methods

2.1. Animal Protocol

The animal protocol used for these studies is a well-established and well-characterized sub-chronic treatment protocol for producing Cd-induced nephrotoxicity in the rat [21]. The animal research was conducted in compliance with the United States NIH Guide for the Care and Use of Laboratory Animals (National Research Council of the National Academies, 2011), and all studies were approved by the Institutional Animal Care and Use Committee of Midwestern University. Adult male Sprague-Dawley rats weighing 250–300 g were purchased from Envigo (Indianapolis, IN, USA). The rats were housed socially with two rats per plastic cage, and the animals were maintained on a 12/12 h light/dark cycle. For the Cd treatment group, animals ($N = 6$) received daily (Monday–Friday) subcutaneous injections of $CdCl_2$ at a Cd dose of 0.6 mg (5.36 µmoles)/kg in 0.25–0.35 mL isotonic saline for 12 weeks, while the vehicle control animals ($N = 6$) received daily injections of isotonic saline. At the end of the 12-week treatment protocol, animals were placed in individual metabolic cages for 24 h to collect urine samples. Animals were allowed free access to water and food ad libitum, with the exception that food was restricted when the animals were in the metabolic cages. Before the start of the dosing regimen, the Cd concentration in the stock solution was confirmed by Chemical Solutions, Inc. (Harrisburg, PA, USA) using the technique of inductively coupled plasma mass spectrometry as previously described [22].

At the end of the protocol, the animals were anesthetized with ketamine/xylazine (67/7 mg/kg) by intraperitoneal injection and euthanized by exsanguination and pneumothorax while under anesthesia. Prior to exsanguination and pneumothorax, the kidneys were removed and processed for RNA isolation as described below.

2.2. Biomarker Determination

The 24 h urine was collected and portioned into 0.5–1.0 mL aliquots. The aliquots were frozen at −80 °C and later assayed for protein, creatinine, and the biomarkers of interest. In some cases, prior to freezing, the urine aliquots were stabilized in proprietary buffers and other reagents that are recommended for MAGPIX-based assays that were used for some of the analyses. The urinary levels of cystatin C, Kim-1, and β_2 microglobulin were determined by microsphere-based Luminex xMAP technology using the MagPix xPONENT 4.1 equipment (Luminex Corp., Austin, TX, USA) The Multiplex technology allows for the determination of multiple analytes in a single sample and provides much greater sensitivities and dynamic ranges than commonly used ELISAs. This technique is similar to the assay that has been used to determine urinary levels of Kim-1 in our previous studies [21,23,24]. Urinary levels of creatinine were determined using a previously described colorimetric method [25]. Urinary protein levels were determined using the Bradford Coomassie blue assay as previously described (Thermo Fisher Scientific, Waltham, MA, USA) [21]. With the dosing protocol used in these studies, Cd-treated animals tended to gain less weight than control animals [21]. Accordingly, all urinary parameters were expressed as units excreted per kg body weight per 24 h.

2.3. RNA Isolation

Renal cortices were removed and snap frozen in liquid nitrogen. Frozen tissues were placed in pre-chilled (−80 °C) RNA*later*®-ICE Frozen Tissue Transition Solution (Invitrogen by Thermo Fisher Scientific) and stored at −80 °C until samples were processed for RNA isolation. Total RNA was isolated from the tissues using the mirVana™ miRNA Isolation Kit (Invitrogen by Thermo Fisher Scientific) following the manufacturer's recommended protocol. The integrity of the RNA samples was evaluated using a Thermo Scientific™ NanoDrop™ 2000 spectrophotometer, (Thermo Fisher Scientific) and by examining 28S and 18S ribosomal RNA bands using denaturing agarose gel electrophoresis and ethidium bromide staining.

2.4. μParaflo™ MicroRNA Microarray Assay

Microarray analysis for rat miRNAs (Sanger miRBase release 21) was performed using μParaflo™ microfluidic chip technology (LC Sciences, Houston, TX, USA). Total RNA (1 μg) from Cd-treated and saline control rats (N = 6 per group) was 3′-extended with a poly(A) tail, and an oligonucleotide was then ligated to the poly(A) tail for subsequent fluorescent dye staining. Hybridization was performed overnight on a μParaflo™ microfluidic chip using a microcirculation pump (Atactic Technologies, Houston, TX, USA) [26,27]. The microfluidic chip contained 761 unique detection probes made by in situ synthesis using Photogenerated Reagent (PGR) chemistry, and the probes consisted of a chemically modified nucleotide segment complementary to a target miRNA from miRbase (http://mirbase.org) or control RNA with a polyethylene glycol spacer segment. The hybridization melting temperatures were balanced by chemical modifications of the detection probes. Hybridization was performed using 100 μL of 6x SSPE buffer (0.90 M NaCl, 60 mM Na_2HPO_4, 6 mM EDTA, pH 6.8) containing 25% formamide at 40 °C. After RNA hybridization, tag-conjugating Alexa Fluor® 647 dye was circulated through the microfluidic chip for dye staining. Fluorescence images were collected using a laser scanner (GenePix 4000B, Molecular Devices, San Jose, CA, USA) and digitized using Array-Pro image analysis software (Version 4.0, Media Cybernetics, Rockville, MD, USA, 1981). Data were analyzed by first subtracting the background and then normalizing the signals using a LOWESS filter (locally weighted regression) [28]. Statistical analysis was performed by LC Sciences using an unpaired *t* test (N = 6 per group, $p \leq 0.05$).

2.5. MicroRNA Real-Time PCR

The cDNA template for PCR was prepared using 10 ng of total RNA sample and the TaqMan® Advanced miRNA cDNA Synthesis kit (Thermo Fisher Scientific) following the manufacturer's recommended protocol. MicroRNA expression in the samples was assessed using TaqMan® Advanced miRNA assays and an Applied Biosystems QuantStudio 5 real-time PCR system. Selected miRNAs that demonstrated a statistically significant ($p \leq 0.05$) altered expression using μParaflo™ microRNA microarray were validated using the following TaqMan® Advanced miRNA assays: miR-21-5p (rno481342_mir), miR-34a-5p (rno481304_mir), miR-146b-5p (rno480941_mir), miR-224-5p (rno481010_mir), miR-3084a-3p (rno481040_mir), miR-3084c-3p (rno481313_mir), and miR-455-3p (rno481396_mir). As an endogenous control, miR-26a-5p (hsa477995_mir) was used since this miRNA was abundantly expressed in the rat renal cortex and the expression levels were not affected with Cd treatment. As an additional validation of the μParaflo™ microRNA microarray assay, the expression of miR-423-5p (rno481159_mir) was examined by real-time PCR since the expression of this miRNA was not significantly altered with Cd treatment. The fold change in miRNA expression between saline control and Cd-treated samples was determined using the comparative CT method as previously described using QuantStudio™ Design Analysis Software (Version 1.4, Applied Biosystems by Thermo Fisher Scientific, Carlsbad, CA, USA, 2016) [29,30]. Statistical analysis was performed on the $2^{-\Delta CT}$ values using an unpaired *t*-test ($N = 6$ per group, $p \leq 0.05$) with GraphPad Prism software (Version 7.00, GraphPad Software Inc., La Jolla, CA, USA, 2016).

3. Results

3.1. Cd-Induced Kidney Injury in a Sub-Chronic Rat Model

The 12-week sub-chronic Cd-treatment protocol resulted in kidney injury. In the Cd-treated animals, there was a significant 4-fold increase in the 24 h urine volume (Figure 1A) and a significant 2.2-fold increase in the 24 h urinary protein excretion (Figure 1B), with no significant change in the urinary creatinine excretion (Figure 1C). There was also a significant 21.7-fold increase in urinary Kim-1 (Figure 1D), a significant 6.1-fold increase in urinary β_2 microglobulin (Figure 1E), and a significant 7.4-fold increase in urinary cystatin C (Figure 1F). These changes in urinary parameters and biomarkers were essentially identical to reported recent studies from the Prozialeck laboratory [21,23,24,31].

3.2. μParaflo™ MicroRNA Microarray

To determine if Cd alters the miRNA profile in the renal cortex of 12-week Cd-treated rats, the miRNA expression profile between Cd-treated and saline control rats was compared using μParaflo™ microRNA microarray analysis. As shown in the heat map in Figure 2, the expression of 54 miRNAs were significantly decreased in the Cd-treated group versus the saline control group, while the expression of 44 miRNAs were significantly increased. More detailed information concerning the miRNAs demonstrating significantly increased expression or decreased expression is shown in Tables 1 and 2, respectively. miR-21-5p, miR-34a-5p, miR-146b-5p, miR-149-3p, miR-224-5p, miR-451-5p, miR-1949, miR-3084a-3p, and miR-3084c-3p demonstrated more abundant expression and a two-fold or greater significant increase in expression with Cd treatment. miR-193b-3p, miR-455-3p, and miR-342-3p demonstrated more abundant expression and a two-fold or greater significant decreased expression with Cd treatment.

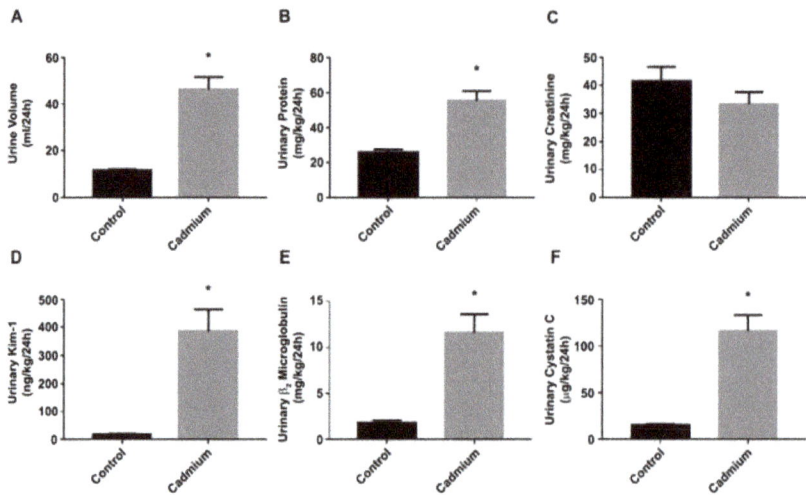

Figure 1. Assessment of Cd-induced nephrotoxicity in a 12-week sub-chronic rat model. Male Sprague-Dawley rats received daily subcutaneous injections of Cd (0.6 mg/kg/day) 5-days a week for 12 weeks, while the controls received injections of isotonic saline. (**A**) Urine volume; (**B**) urinary protein; (**C**) urinary creatinine; (**D**) urinary Kim-1; (**E**) urinary β_2 microglobulin; (**F**) urinary cystatin C. The data are mean $\pm$ SEM; an asterisk (*) indicates statistical significance from the saline control ($N = 6$ per group, unpaired t-test, $p \leq 0.05$).

3.3. Real-Time PCR Validation

Real-time PCR was used to validate the expression of selected miRNAs that demonstrated more abundant expression and demonstrated a two-fold or greater change in expression as determined by the µParaflo™ microRNA microarray analysis. As shown in Figure 3, real-time PCR demonstrated a significant increase in the Cd-treated group for the following miRNAs: a 2.7-fold increase in miR-21-5p (Figure 3A), a 10.8-fold increase in miR-34a-5p (Figure 3B), a 2-fold increase in miR-146b-5p (Figure 3C), a 10.2-fold increase in miR-224-5p (Figure 3D), a 2.4-fold increase in miR-3084a-3p (Figure 3E), and a 3.3-fold increase in miR-3084c-3p (Figure 3F). By contrast, real-time PCR validation demonstrated a significant 52% decrease in miR-455-3p expression in the Cd-treatment group (Figure 3G). As a control, we also examined the expression of a miRNA (miR-423-5p) that did not demonstrate altered expression in the Cd-treatment group versus the saline control group according to the microarray analysis, and real-time PCR analysis confirmed there was no significant difference in the expression of miR-423-5p between the saline control and Cd-treatment group (Figure 3H).

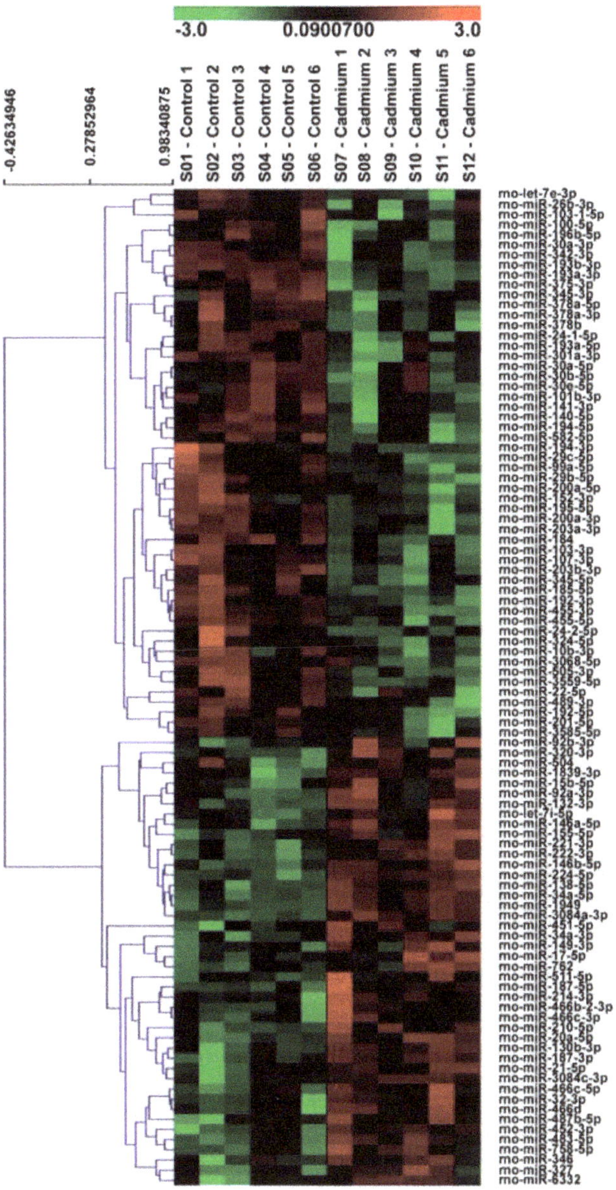

Figure 2. Effects of Cd on miRNA expression in the rat renal cortex. Microarray heat map demonstrating significant differences in the expression of miRNAs in the renal cortex of Cd-treated (0.6 mg/kg/day, 5 days per week for 12 weeks) male Sprague-Dawley rats versus saline controls. Cadmium significantly decreased the expression of 54 miRNAs and increased the expression of 44 miRNAs ($N = 6$ per group, unpaired *t*-test, $p \leq 0.05$).

Table 1. MicroRNAs with significantly increased expression in the renal cortex of Cd-treated rats as determined by µParaflo™ microRNA microarray analysis.

MicroRNA	*p*-Value	Control Mean (RFS *)	Cadmium Mean (RFS *)	Log 2 (Cadmium/Control)
miR-3084a-3p	1.05×10^{-6}	1019	3016	1.57
miR-34a-5p	4.57×10^{-6}	99	612	2.62
miR-1949	1.10×10^{-5}	41	326	2.98
miR-224-5p	3.75×10^{-5}	12	390	5.06
miR-222-3p	3.00×10^{-4}	622	1127	0.86
miR-221-3p	3.95×10^{-4}	968	1643	0.76
miR-146b-5p	8.79×10^{-4}	200	558	1.48
miR-210-5p	1.81×10^{-3}	1140	1740	0.61
miR-20a-5p	1.87×10^{-3}	1179	1756	0.58
miR-146a-5p	2.89×10^{-3}	3840	5884	0.62
miR-3084c-3p	4.34×10^{-3}	1174	3419	1.54
miR-92a-3p	6.52×10^{-3}	1083	1926	0.83
miR-21-5p	6.98×10^{-3}	10,943	22,388	1.03
miR-466b-2-3p	7.25×10^{-3}	2101	3143	0.58
miR-320-3p	1.18×10^{-2}	1377	1882	0.45
miR-15b-5p	1.29×10^{-2}	1032	1647	0.67
miR-466c-3p	1.29×10^{-2}	3427	5220	0.61
miR-214-3p	1.64×10^{-2}	1582	2094	0.40
miR-483-5p	1.74×10^{-2}	711	1184	0.74
miR-149-3p	1.78×10^{-2}	1573	3796	1.27
let-7i-5p	2.67×10^{-2}	3498	4619	0.40
miR-762	2.84×10^{-2}	915	1702	0.90
miR-466d	3.47×10^{-2}	370	675	0.87
miR-346	3.57×10^{-2}	315	440	0.48
miR-17-5p	3.60×10^{-2}	877	1269	0.53
miR-451-5p	3.63×10^{-2}	552	1177	1.09
miR-92b-3p	3.81×10^{-2}	471	759	0.69
miR-466c-5p	3.83×10^{-2}	229	389	0.76
miR-32-3p	4.07×10^{-2}	622	1144	0.88
Statistically significant transcripts with low signals (signal < 500)				
miR-138-5p	4.00×10^{-4}	43	140	1.71
miR-130b-3p	7.36×10^{-4}	12	59	2.25
miR-187-3p	3.82×10^{-3}	84	242	1.53
miR-155-5p	6.57×10^{-3}	33	197	2.57
miR-1839-3p	7.09×10^{-3}	293	417	0.51
miR-187-5p	8.23×10^{-3}	66	114	0.79
miR-132-3p	9.57×10^{-3}	59	189	1.66
miR-34a-3p	1.08×10^{-2}	7	24	1.86
miR-452-3p	1.71×10^{-2}	5	27	2.36
miR-511-5p	1.99×10^{-2}	36	90	1.32
miR-758-5p	2.08×10^{-2}	203	281	0.47
miR-487b-5p	2.13×10^{-2}	29	69	1.22
miR-327	2.19×10^{-2}	28	51	0.84
miR-504	4.10×10^{-2}	98	136	0.47
miR-6332	4.30×10^{-2}	16	28	0.82

* Relative fluorescent signal.

Table 2. MicroRNAs with significantly decreased expression in the renal cortex of Cd-treated rats as determined by µParaflo™ microRNA microarray analysis.

MicroRNA	*p*-Value	Control Mean (RFS *)	Cadmium Mean (RFS *)	Log 2 (Cadmium/Control)
miR-193b-3p	2.29×10^{-5}	445	137	−1.70
miR-185-5p	2.81×10^{-5}	1150	628	−0.87
miR-455-3p	2.06×10^{-4}	764	258	−1.57
miR-195-5p	4.76×10^{-4}	4374	3035	−0.53
miR-200a-3p	2.31×10^{-3}	5998	3725	−0.69
miR-101b-3p	2.56×10^{-3}	465	285	−0.71
miR-194-5p	2.72×10^{-3}	13,390	7697	−0.80
miR-99a-5p	2.79×10^{-3}	5468	3596	−0.60
miR-505-3p	3.59×10^{-3}	539	371	−0.54
miR-342-3p	4.25×10^{-3}	1871	845	−1.15
miR-203a-3p	5.21×10^{-3}	1327	730	−0.86
miR-378a-3p	6.43×10^{-3}	2576	1616	−0.67
miR-378a-5p	6.67×10^{-3}	416	233	−0.83
miR-140-5p	7.56×10^{-3}	403	228	−0.82
miR-378b	9.43×10^{-3}	1985	1298	−0.61
miR-103-3p	1.73×10^{-2}	2717	2000	−0.44
miR-107-3p	1.74×10^{-2}	2781	2052	−0.44
miR-192-5p	2.31×10^{-2}	13,962	11,183	−0.32
miR-152-3p	2.98×10^{-2}	971	683	−0.51
miR-100-5p	3.39×10^{-2}	2133	1318	−0.70
miR-30a-3p	3.73×10^{-2}	837	552	−0.60
miR-30a-5p	3.81×10^{-2}	15,805	12,197	−0.37
miR-22-5p	3.84×10^{-2}	939	812	−0.21
miR-30b-5p	3.93×10^{-2}	14,704	11,704	−0.33
miR-196b-5p	4.03×10^{-2}	464	318	−0.54
miR-489-3p	4.21×10^{-2}	485	311	−0.64
miR-30e-5p	4.68×10^{-2}	10,074	6429	−0.65
Statistically significant transcripts with low signals (signal < 500)				
miR-203b-3p	6.03×10^{-5}	146	31	−2.25
miR-192-3p	7.66×10^{-5}	299	105	−1.50
miR-193a-3p	2.05×10^{-4}	328	104	−1.65
miR-455-5p	3.55×10^{-4}	70	17	−2.05
miR-184	6.06×10^{-4}	27	5	−2.52
miR-375-3p	7.44×10^{-4}	39	11	−1.86
miR-345-5p	1.04×10^{-3}	183	103	−0.82
miR-29b-5p	2.03×10^{-3}	148	78	−0.92
miR-301a-3p	3.09×10^{-3}	122	60	−1.03
miR-3559-5p	5.48×10^{-3}	298	161	−0.89
miR-582-5p	9.16×10^{-3}	165	99	−0.73
miR-345-3p	9.25×10^{-3}	58	36	−0.70
miR-24-1-5p	1.07×10^{-2}	98	53	−0.88
miR-29c-5p	1.07×10^{-2}	274	161	−0.77
miR-24-2-5p	1.12×10^{-2}	276	181	−0.61
miR-10b-3p	1.54×10^{-2}	200	121	−0.72
miR-3068-5p	1.86×10^{-2}	162	113	−0.52
miR-200a-5p	1.87×10^{-2}	133	73	−0.86
miR-201-5p	2.26×10^{-2}	67	33	−1.00
miR-141-3p	2.41×10^{-2}	171	83	−1.05
miR-194-3p	2.63×10^{-2}	83	44	−0.92
miR-324-5p	2.73×10^{-2}	243	180	−0.43
miR-26b-3p	3.38×10^{-2}	27	10	−1.47
miR-193a-5p	3.45×10^{-2}	20	5	−2.12
miR-3585-5p	3.50×10^{-2}	67	34	−0.98
let-7e-3p	4.06×10^{-2}	47	23	−1.00
miR-103-1-5p	4.71×10^{-2}	32	22	−0.52

* Relative fluorescent signal.

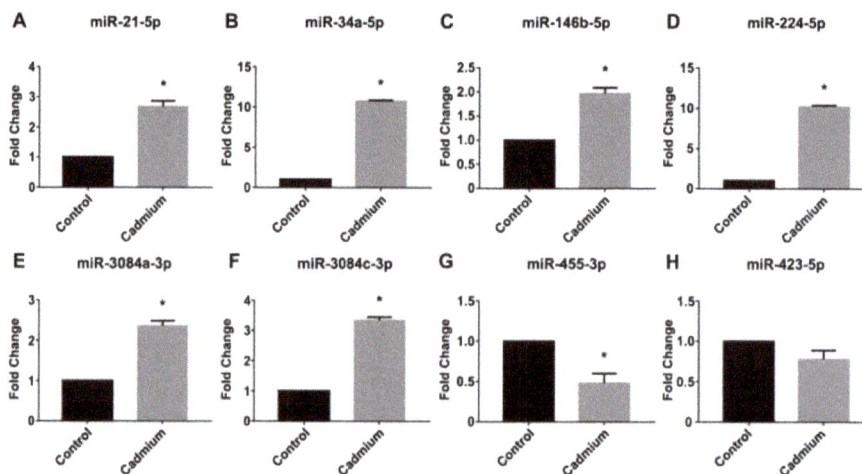

Figure 3. Real-time PCR validation of Cd-dysregulated miRNAs. TaqMan® Advanced miRNA assays were used to validate selected Cd-dysregulated miRNAs. (**A**) miR-21-5p; (**B**) miR-34a-5p; (**C**) miR-146b-5p; (**D**) miR-224-5p; (**E**) miR-3084a-3p; (**F**) miR-3084c-3p; (**G**) miR-455-3p; (**H**) miR-423-5p. The comparative CT method was used to determine the fold change (±SEM), and an asterisk (*) indicates a statistically significant change in expression in the Cd-treated group versus the saline control ($N = 6$ per group, unpaired t-test, $p \leq 0.05$).

4. Discussion

The purpose of this research was to determine if Cd-induced nephrotoxicity is associated with dysregulated miRNA expression in the renal cortex. As a research model of Cd-induced nephrotoxicity, we used a well-characterized sub-chronic in vivo research model in which male Sprague-Dawley rats received daily subcutaneous injections of $CdCl_2$ (0.6 mg/kg/day, 5 days per week for 12 weeks) [21]. The Cd-induced damage to the proximal tubule was confirmed by demonstrating polyuria and proteinuria in the Cd-treatment group versus the saline control group [21]. The fact that the 12-week Cd-treated animals developed polyuria and proteinuria without a significant change in creatinine excretion supports the fact that the Cd-induced kidney injury is at the level of the proximal tubule [5,6,21,24,31,32]. Some of the urinary biomarkers that have been used to monitor kidney injury include Kim-1, β_2 microglobulin, and cystatin C [21,31,33–36]. All three of these urinary biomarkers have previously been shown to be elevated in the animal model of Cd-induced kidney injury used for this study, and consistent with these previous findings, all three biomarkers were significantly increased in our study [21,31].

The microarray results demonstrated a significant increase in the expression of 44 miRNAs in the renal cortices from the Cd-treated animals versus the saline control group. miR-21-5p, miR-34a-5p, miR-146b-5p, miR-149-3p, miR-224-5p, miR-451-5p, miR-1949, miR-3084a-3p, and miR-3084c-3p were more abundantly expressed and demonstrated a two-fold or greater increased expression in the Cd-treatment group. We used real-time PCR to confirm the increased expression of miR-21-5p, miR-34a-5p, mir-146b-5p, miR-224-5p, miR-3084a-3p, and miR-3084c-3p in the renal cortices from Cd-treated animals versus the saline controls. The elevated level of some of these miRNAs have also been demonstrated in other models of kidney injury. Previous research demonstrated increased expression of miR-21 in the kidneys of mice with ischemia/reperfusion injury or gentamicin-induced kidney injury, and miR-21 levels were also increased in the urine of patients with acute kidney injury versus healthy patients [37]. The cellular effects of elevated miR-21 on proximal tubule epithelial cells may be both protective and/or damaging as miR-21 has been shown to limit damage resulting from reactive oxygen species and affect apoptosis and fibrosis [20,38,39]. Several rodent studies have demonstrated that drug-induced kidney injury is associated with increased expression of p53-responsive miR-34a, and this miRNA has been shown to suppress autophagy in tubular epithelial cells in a mouse model of ischemia/reperfusion-induced acute kidney injury [40–43]. The expression of miR-146-5p was shown to be increased with fibrosis in a mouse model of folic-acid-induced kidney injury, and in mouse models of ischemia/reperfusion injury and unilateral urethral obstruction-induced fibrosis [44]. Additionally, miR-146-5p expression was increased in human renal cortices with documented severe kidney injury or fibrosis [44]. Although not directly linked to kidney injury, upregulation of miR-149-3p may decrease clonogenicity and induce apoptosis by targeting polo-like kinase 1 (PLK1) [45]. Additionally, miR-149-3p inhibits the proliferation, migration, and invasion of bladder cancer cells by targeting the S100A4 protein, which is involved with cellular differentiation, motility, and regulating transcription [46]. Upregulation of miR-224-5p has been found to occur as part of the protective adaptive response of hepatocytes during acetaminophen-induced toxicity, and this upregulation of miR-224-5p was associated with suppression of drug metabolizing enzyme levels [47]. Using a streptozotocin-induced diabetic rat model, Mohan et al. demonstrated the utility of urinary exosomal miR-451-5p as an early biomarker of diabetes-associated nephropathy, and the elevated levels of miR-451-5p appeared to be protective against kidney fibrosis [48]. Increased expression of miR-1949 was previously shown in injured kidney tissue in a rat model of deep hypothermic circulatory arrest [49]. Previous research demonstrated increased expression of miR-3084-3p in the renal cortices of mice treated with ^{177}Lu-octreotate, and this radionuclide therapy used for treating neuroendocrine tumors is known to cause renal toxicity at the level of the proximal tubule [50,51].

Although it did not meet the criteria of two-fold or greater increased expression, miR-320-3p demonstrated significantly increased expression in the Cd-treatment group. Previous research has identified β-catenin mRNA as a target of miR-320 [52]. This is relevant because our research group demonstrated using both in vitro and in vivo research models that prior to inducing cell death, Cd disrupts cadherin-dependent cell–cell adhesions with a resulting loss of cadherin and β-catenin at cell–cell contacts [6,53–57].

The microarray results demonstrated significantly decreased expression of 54 miRNAs in the renal cortices from the Cd-treated group versus the saline control group, and miR-193b-3p, miR-455-3p, and miR-342-3p demonstrated a more abundant expression level and a two-fold or greater decreased expression in the Cd-treatment group versus the saline control group. As shown here, the decreased expression of miR-455-3p in the renal cortices from Cd-treatment animals was confirmed by real-time PCR. Fabbri et al. treated HepG2 human hepatoma cells with 10 μM Cd for 24 h and reported decreased expression of 12 miRNAs, including members of the let-7 family (let-7a, let-7b, let-7e, and let-7g) and miR-455-3p [58]. The microarray results presented here also demonstrated decreased expression of let-7e-3p, and both the microarray and real-time PCR validation from our study demonstrated decreased expression of miR-455-3p. The top pathways that were identified to be affected by the altered miRNA expression profile in the Cd-treated HepG2 cells were focal adhesion and the MAPK signaling pathway, and members of the let-7 miRNA family are known to serve a tumor suppressor role [58,59]. Another study used lentiviruses to silence β-catenin in gastric cancer cell lines and found the dysregulated expression of a number of miRNAs including increased expression of miR-210 and decreased expression of miR-455-3p [60]. Both miR-210-5p and miR-455-3p were also dysregulated in our study, and we have previously shown that Cd alters β-catenin sub-cellular localization and activity in NRK-52E rat proximal tubule epithelial cells and causes a loss of β-catenin at cell–cell contacts [6,53–57]. miRNA-193b-3p was found to be downregulated in the renal cortices from Cd-treated animals, and downregulation of this miRNA has been shown to promote autophagy and cell survival in mouse motor neuron-like cells [61]. Additionally, downregulated miR-342-3p expression is part of a shared dysregulated miRNA expression profile in both mutated non-invasive follicular thyroid neoplasms with papillary-like nuclear features (NIFTPs) and infiltrative and invasive follicular variants of papillary thyroid carcinomas (FVPTCs); and in contrast to wild type NIFTPs, this dysregulated miRNA expression profile is predicted to promote an invasive-like phenotype by altering cell adhesion and cell migration pathways [62].

It is important to note that the sub-chronic animal model of Cd-induced kidney injury that was employed for the present studies is similar to protocols that have been used by other investigators [24]. With this sub-chronic Cd-treatment protocol, the classic symptoms of polyuria and proteinuria without a change in creatinine excretion appear after 9–10 weeks of the Cd-dosing protocol. Even after 12 weeks, the level of injury in the proximal tubule is relatively mild; as a result, this animal model is useful for identifying cellular alterations and biomarkers at early stages of Cd-induced proximal tubule injury prior to overt dysfunction. The fact that changes in miRNA expression are quite pronounced at 12 weeks, when the level of injury is mild, suggests that the changes in miRNA expression represent early events in the pathophysiology of Cd-induced kidney injury. More detailed time course studies, beyond the scope of this report, will be needed to clarify this issue. An important consideration for this research is the tissue source of the Cd-dysregulated miRNAs, as the renal cortices were obtained from non-perfused animals. In our experience with this animal model, there is limited blood associated with the isolated renal cortices compared to the amount of renal tissue. However, to confirm the renal source of the miRNAs, future studies will be performed using in situ hybridization.

5. Conclusions

The results of these studies demonstrate that Cd-induced nephrotoxicity is associated with dysregulated miRNA expression in the rat renal cortex. These dysregulated miRNAs may serve a protective role and/or promote injury of the Cd-exposed proximal tubule epithelial cells. Identifying the mRNA targets of these dysregulated miRNAs and the associated cellular signaling pathways may help to identify novel therapeutic strategies for preventing and treating kidney disease. In addition to blocking gene expression at the post-transcriptional level, these Cd-dysregulated miRNAs may also serve as useful non-invasive urinary biomarkers of Cd exposure or Cd-induced kidney injury. Future studies from our research group will determine if the expression of these miRNAs are dysregulated in the renal cortex prior to the Cd-induced kidney injury that is seen after 12 weeks of Cd treatment (0.6 mg/kg/day, 5 days per week), and can be used as non-invasive urinary biomarkers of Cd-induced kidney injury. Finally, we will utilize bioinformatics and an in vitro research model to determine the mRNA targets and cellular effects of these dysregulated miRNAs through the use of both miRNA mimics and inhibitors. These studies will provide useful information regarding the molecular mechanisms of Cd-induced damage to proximal tubule epithelial cells, and will determine if miRNAs can be used as early non-invasive biomarkers of Cd-induced damage to the proximal tubule.

Acknowledgments: The research presented in this manuscript was supported by Midwestern University intramural research funds and by student research funds from the Biomedical Sciences Program. The authors thank Catherine Lencioni for assistance with EndNote.

Author Contributions: M.J.F., J.R.E., and W.C.P. conceived and designed the experiments, interpreted results, assisted with the animal study, and wrote the paper. L.A.C.A. assisted with the animal study, performed RNA isolation and real-time PCR studies, performed statistical analyses, and prepared figures. D.R., R.S., R.P., A.P., and S.Z. assisted with the animal study, RNA isolation, and real-time PCR. P.C.L. assisted with the animal study and performed urinary biomarker analysis. Z.S.-R., A.W., N.P., and A.R. assisted with RNA isolation and real-time PCR.

References

1. ATSDR's Substance Priority List. Available online: https://www.atsdr.cdc.gov/spl/index.html (accessed on 30 January 2018).
2. Jarup, L.; Akesson, A. Current status of cadmium as an environmental health problem. *Toxicol. Appl. Pharmacol.* **2009**, *238*, 201–208. [CrossRef] [PubMed]
3. Bridges, C.C.; Zalups, R.K. Molecular and ionic mimicry and the transport of toxic metals. *Toxicol. Appl. Pharmacol.* **2005**, *204*, 274–308. [CrossRef] [PubMed]
4. Jarup, L. Cadmium overload and toxicity. *Nephrol. Dial. Transplant. Off. Publ. Eur. Dial. Transpl. Assoc. Eur. Renal Assoc.* **2002**, *17*, 35–39. [CrossRef]

5. Kjellstrom, T. Renal effects. In *Cadmium and Health: A Toxicoloogical and Epidemiological Appraisal*; Friberg, L., Elinder, C.-G., Kjellstrom, T., Nordberg, G.F., Eds.; CRC Press: Boca Raton, FL, USA, 1986; Volume 2, pp. 21–109.

6. Prozialeck, W.C.; Edwards, J.R. Mechanisms of cadmium-induced proximal tubule injury: New insights with implications for biomonitoring and therapeutic interventions. *J. Pharmacol. Exp. Ther.* **2012**, *343*, 2–12. [CrossRef] [PubMed]

7. Thevenod, F.; Lee, W.K. Cadmium and cellular signaling cascades: Interactions between cell death and survival pathways. *Arch. Toxicol.* **2013**, *87*, 1743–1786. [CrossRef] [PubMed]

8. Ambros, V. MicroRNAs: Tiny regulators with great potential. *Cell* **2001**, *107*, 823–826. [CrossRef]

9. Bartel, D.P. MicroRNAs: Genomics, biogenesis, mechanism, and function. *Cell* **2004**, *116*, 281–297. [CrossRef]

10. Filipowicz, W.; Bhattacharyya, S.N.; Sonenberg, N. Mechanisms of post-transcriptional regulation by microRNAs: Are the answers in sight? *Nat. Rev. Genet.* **2008**, *9*, 102–114. [CrossRef] [PubMed]

11. Krol, J.; Loedige, I.; Filipowicz, W. The widespread regulation of microRNA biogenesis, function and decay. *Nat. Rev. Genet.* **2010**, *11*, 597–610. [CrossRef] [PubMed]

12. Chandrasekaran, K.; Karolina, D.S.; Sepramaniam, S.; Armugam, A.; Wintour, E.M.; Bertram, J.F.; Jeyaseelan, K. Role of microRNAs in kidney homeostasis and disease. *Kidney Int.* **2012**, *81*, 617–627. [CrossRef] [PubMed]

13. Ho, J.; Kreidberg, J.A. MicroRNAs in renal development. *Pediatr. Nephrol.* **2013**, *28*, 219–225. [CrossRef] [PubMed]

14. Papadopoulos, T.; Belliere, J.; Bascands, J.L.; Neau, E.; Klein, J.; Schanstra, J.P. MiRNAs in urine: A mirror image of kidney disease? *Expert Rev. Mol. Diagn.* **2015**, *15*, 361–374. [CrossRef] [PubMed]

15. Trionfini, P.; Benigni, A.; Remuzzi, G. MicroRNAs in kidney physiology and disease. *Nat. Rev. Nephrol.* **2015**, *11*, 23–33. [CrossRef] [PubMed]

16. Zhou, P.; Chen, Z.; Zou, Y.; Wan, X. Roles of non-coding RNAs in acute kidney injury. *Kidney Blood Press. Res.* **2016**, *41*, 757–769. [CrossRef] [PubMed]

17. Badal, S.S.; Danesh, F.R. MicroRNAs and their applications in kidney diseases. *Pediatr. Nephrol.* **2015**, *30*, 727–740. [CrossRef] [PubMed]

18. Fan, P.C.; Chen, C.C.; Chen, Y.C.; Chang, Y.S.; Chu, P.H. MicroRNAs in acute kidney injury. *Hum. Genom.* **2016**, *10*, 29. [CrossRef] [PubMed]

19. Gerlach, C.V.; Vaidya, V.S. MicroRNAs in injury and repair. *Arch. Toxicol.* **2017**, *91*, 2781–2797. [CrossRef] [PubMed]

20. Pavkovic, M.; Vaidya, V.S. MicroRNAs and drug-induced kidney injury. *Pharmacol. Ther.* **2016**, *163*, 48–57. [CrossRef] [PubMed]

21. Prozialeck, W.C.; Vaidya, V.S.; Liu, J.; Waalkes, M.P.; Edwards, J.R.; Lamar, P.C.; Bernard, A.M.; Dumont, X.; Bonventre, J.V. Kidney injury molecule-1 is an early biomarker of cadmium nephrotoxicity. *Kidney Int.* **2007**, *72*, 985–993. [CrossRef] [PubMed]

22. Prozialeck, W.C.; Lamar, P.C.; Edwards, J.R. Effects of sub-chronic Cd exposure on levels of copper, selenium, zinc, iron and other essential metals in rat renal cortex. *Toxicol. Rep.* **2016**, *3*, 740–746. [CrossRef] [PubMed]

23. Prozialeck, W.C.; Edwards, J.R.; Vaidya, V.S.; Bonventre, J.V. Preclinical evaluation of novel urinary biomarkers of cadmium nephrotoxicity. *Toxicol. Appl. Pharmacol.* **2009**, *238*, 301–305. [CrossRef] [PubMed]

24. Prozialeck, W.C.; Edwards, J.R.; Lamar, P.C.; Liu, J.; Vaidya, V.S.; Bonventre, J.V. Expression of kidney injury molecule-1 (Kim-1) in relation to necrosis and apoptosis during the early stages of Cd-induced proximal tubule injury. *Toxicol. Appl. Pharmacol.* **2009**, *238*, 306–314. [CrossRef] [PubMed]

25. Shoucri, R.M.; Pouliot, M. Some observations on the kinetics of the Jaffe reaction for creatinine. *Clin. Chem.* **1977**, *23*, 1527–1530. [PubMed]

26. Gao, X.; Gulari, E.; Zhou, X. In situ synthesis of oligonucleotide microarrays. *Biopolymers* **2004**, *73*, 579–596. [CrossRef] [PubMed]

27. Zhu, Q.; Hong, A.; Sheng, N.; Zhang, X.; Matejko, A.; Jun, K.Y.; Srivannavit, O.; Gulari, E.; Gao, X.; Zhou, X. Microparaflo biochip for nucleic acid and protein analysis. *Methods Mol. Biol.* **2007**, *382*, 287–312. [PubMed]

28. Bolstad, B.M.; Irizarry, R.A.; Astrand, M.; Speed, T.P. A comparison of normalization methods for high density oligonucleotide array data based on variance and bias. *Bioinformatics* **2003**, *19*, 185–193. [CrossRef] [PubMed]

29. Baxter, S.S.; Carlson, L.A.; Mayer, A.M.; Hall, M.L.; Fay, M.J. Granulocytic differentiation of HL-60 promyelocytic leukemia cells is associated with increased expression of Cul5. *In Vitro Cell. Dev. Biol. Anim.* **2009**, *45*, 264–274. [CrossRef] [PubMed]

30. Schmittgen, T.D.; Livak, K.J. Analyzing real-time PCR data by the comparative C(T) method. *Nat. Protoc.* **2008**, *3*, 1101–1108. [CrossRef] [PubMed]

31. Prozialeck, W.C.; VanDreel, A.; Ackerman, C.D.; Stock, I.; Papaeliou, A.; Yasmine, C.; Wilson, K.; Lamar, P.C.; Sears, V.L.; Gasiorowski, J.Z.; et al. Evaluation of cystatin C as an early biomarker of cadmium nephrotoxicity in the rat. *Biometals Int. J. Role Metal Ions Biol. Biochem. Med.* **2016**, *29*, 131–146. [CrossRef] [PubMed]

32. Piscator, M. The nephropathy of chronic cadmium poisoning. In *Cadmium, Handbook of Experimental Pharmacology*; Faulkes, E.C., Ed.; Springer: New York, NY, USA, 1986; Volume 80, pp. 194–197.

33. Bernard, A. Renal dysfunction induced by cadmium: Biomarkers of critical effects. *Biometals Int. J. Role Metal Ions Biol. Biochem. Med.* **2004**, *17*, 519–523. [CrossRef]

34. Lauwerys, R.R.; Bernard, A.M.; Roels, H.A.; Buchet, J.P. Cadmium: Exposure markers as predictors of nephrotoxic effects. *Clin. Chem.* **1994**, *40*, 1391–1394. [PubMed]

35. Kobayashi, E.; Suwazono, Y.; Uetani, M.; Inaba, T.; Oishi, M.; Kido, T.; Nishijo, M.; Nakagawa, H.; Nogawa, K. Estimation of benchmark dose as the threshold levels of urinary cadmium, based on excretion of total protein, beta2-microglobulin, and N-acetyl-beta-D-glucosaminidase in cadmium nonpolluted regions in Japan. *Environ. Res.* **2006**, *101*, 401–406. [CrossRef] [PubMed]

36. Prozialeck, W.C.; Edwards, J.R. Early biomarkers of cadmium exposure and nephrotoxicity. *Biometals Int. J. Role Metal Ions Biol. Biochem. Med.* **2010**, *23*, 793–809. [CrossRef] [PubMed]

37. Saikumar, J.; Hoffmann, D.; Kim, T.M.; Gonzalez, V.R.; Zhang, Q.; Goering, P.L.; Brown, R.P.; Bijol, V.; Park, P.J.; Waikar, S.S.; et al. Expression, circulation, and excretion profile of microRNA-21, -155, and -18a following acute kidney injury. *Toxicol. Sci. Off. J. Soc. Toxicol.* **2012**, *129*, 256–267. [CrossRef] [PubMed]

38. Cheng, Y.; Liu, X.; Zhang, S.; Lin, Y.; Yang, J.; Zhang, C. MicroRNA-21 protects against the H_2O_2-induced injury on cardiac myocytes via its target gene PDCD4. *J. Mol. Cell. Cardiol.* **2009**, *47*, 5–14. [CrossRef] [PubMed]

39. Li, Y.F.; Jing, Y.; Hao, J.; Frankfort, N.C.; Zhou, X.; Shen, B.; Liu, X.; Wang, L.; Li, R. MicroRNA-21 in the pathogenesis of acute kidney injury. *Protein Cell* **2013**, *4*, 813–819. [CrossRef] [PubMed]

40. Bhatt, K.; Zhou, L.; Mi, Q.S.; Huang, S.; She, J.X.; Dong, Z. MicroRNA-34a is induced via p53 during cisplatin nephrotoxicity and contributes to cell survival. *Mol. Med.* **2010**, *16*, 409–416. [CrossRef] [PubMed]

41. Lee, C.G.; Kim, J.G.; Kim, H.J.; Kwon, H.K.; Cho, I.J.; Choi, D.W.; Lee, W.H.; Kim, W.D.; Hwang, S.J.; Choi, S.; et al. Discovery of an integrative network of microRNAs and transcriptomics changes for acute kidney injury. *Kidney Int.* **2014**, *86*, 943–953. [CrossRef] [PubMed]

42. Pavkovic, M.; Riefke, B.; Ellinger-Ziegelbauer, H. Urinary microRNA profiling for identification of biomarkers after cisplatin-induced kidney injury. *Toxicology* **2014**, *324*, 147–157. [CrossRef] [PubMed]

43. Liu, X.J.; Hong, Q.; Wang, Z.; Yu, Y.Y.; Zou, X.; Xu, L.H. MicroRNA-34a suppresses autophagy in tubular epithelial cells in acute kidney injury. *Am. J. Nephrol.* **2015**, *42*, 168–175. [CrossRef] [PubMed]

44. Pellegrini, K.L.; Gerlach, C.V.; Craciun, F.L.; Ramachandran, K.; Bijol, V.; Kissick, H.T.; Vaidya, V.S. Application of small RNA sequencing to identify microRNAs in acute kidney injury and fibrosis. *Toxicol. Appl. Pharmacol.* **2016**, *312*, 42–52. [CrossRef] [PubMed]

45. Shin, C.H.; Lee, H.; Kim, H.R.; Choi, K.H.; Joung, J.G.; Kim, H.H. Regulation of PLK1 through competition between hnRNPK, miR-149-3p and miR-193b-5p. *Cell Death Differ.* **2017**, *24*, 1861–1871. [CrossRef] [PubMed]

46. Yang, D.; Du, G.; Xu, A.; Xi, X.; Li, D. Expression of miR-149-3p inhibits proliferation, migration, and invasion of bladder cancer by targeting S100A4. *Am. J. Cancer Res.* **2017**, *7*, 2209–2219. [PubMed]

47. Yu, D.; Wu, L.; Gill, P.; Tolleson, W.H.; Chen, S.; Sun, J.; Knox, B.; Jin, Y.; Xiao, W.; Hong, H.; et al. Multiple microRNAs function as self-protective modules in acetaminophen-induced hepatotoxicity in humans. *Arch. Toxicol.* **2018**, *92*, 845–858. [CrossRef] [PubMed]

48. Mohan, A.; Singh, R.S.; Kumari, M.; Garg, D.; Upadhyay, A.; Ecelbarger, C.M.; Tripathy, S.; Tiwari, S. Urinary exosomal microRNA-451-5p is a potential early biomarker of diabetic nephropathy in rats. *PLoS ONE* **2016**, *11*, e0154055. [CrossRef] [PubMed]

49. Yu, L.; Gu, T.; Shi, E.; Wang, Y.; Fang, Q.; Wang, C. Dysregulation of renal microRNA expression after deep hypothermic circulatory arrest in rats. *Eur. J. Cardio-Thorac. Surg. Off. J. Eur. Assoc. Cardio-Thorac. Surg.* **2016**, *49*, 1725–1731. [CrossRef] [PubMed]

50. Svensson, J.; Molne, J.; Forssell-Aronsson, E.; Konijnenberg, M.; Bernhardt, P. Nephrotoxicity profiles and threshold dose values for [177Lu]-DOTATATE in nude mice. *Nucl. Med. Biol.* **2012**, *39*, 756–762. [CrossRef] [PubMed]
51. Schuler, E.; Parris, T.Z.; Helou, K.; Forssell-Aronsson, E. Distinct microRNA expression profiles in mouse renal cortical tissue after 177Lu-octreotate administration. *PLoS ONE* **2014**, *9*, e112645. [CrossRef] [PubMed]
52. Hsieh, I.S.; Chang, K.C.; Tsai, Y.T.; Ke, J.Y.; Lu, P.J.; Lee, K.H.; Yeh, S.D.; Hong, T.M.; Chen, Y.L. MicroRNA-320 suppresses the stem cell-like characteristics of prostate cancer cells by downregulating the Wnt/beta-catenin signaling pathway. *Carcinogenesis* **2013**, *34*, 530–538. [CrossRef] [PubMed]
53. Prozialeck, W.C.; Niewenhuis, R.J. Cadmium (Cd^{2+}) disrupts intercellular junctions and actin filaments in LLC-PK1 cells. *Toxicol. Appl. Pharmacol.* **1991**, *107*, 81–97. [CrossRef]
54. Prozialeck, W.C.; Niewenhuis, R.J. Cadmium (Cd^{2+}) disrupts Ca($^{2+}$)-dependent cell-cell junctions and alters the pattern of E-cadherin immunofluorescence in LLC-PK1 cells. *Biochem. Biophys. Res. Commun.* **1991**, *181*, 1118–1124. [CrossRef]
55. Prozialeck, W.C.; Lamar, P.C. Cadmium (Cd^{2+}) disrupts E-cadherin-dependent cell-cell junctions in MDCK cells. *In Vitro Cell. Dev. Biol. Anim.* **1997**, *33*, 516–526. [CrossRef] [PubMed]
56. Prozialeck, W.C.; Lamar, P.C.; Lynch, S.M. Cadmium alters the localization of N-cadherin, E-cadherin, and beta-catenin in the proximal tubule epithelium. *Toxicol. Appl. Pharmacol.* **2003**, *189*, 180–195. [CrossRef]
57. Edwards, J.R.; Kolman, K.; Lamar, P.C.; Chandar, N.; Fay, M.J.; Prozialeck, W.C. Effects of cadmium on the sub-cellular localization of beta-catenin and beta-catenin-regulated gene expression in NRK-52E cells. *Biometals Int. J. Role Metal Ions Biol. Biochem. Med.* **2013**, *26*, 33–42. [CrossRef] [PubMed]
58. Fabbri, M.; Urani, C.; Sacco, M.G.; Procaccianti, C.; Gribaldo, L. Whole genome analysis and microRNAs regulation in HepG2 cells exposed to cadmium. *Altex* **2012**, *29*, 173–182. [CrossRef] [PubMed]
59. Boyerinas, B.; Park, S.M.; Hau, A.; Murmann, A.E.; Peter, M.E. The role of let-7 in cell differentiation and cancer. *Endocr.-Relat. Cancer* **2010**, *17*, F19–F36. [CrossRef] [PubMed]
60. Dong, L.; Deng, J.; Sun, Z.M.; Pan, A.P.; Xiang, X.J.; Zhang, L.; Yu, F.; Chen, J.; Sun, Z.; Feng, M.; et al. Interference with the beta-catenin gene in gastric cancer induces changes to the miRNA expression profile. *Tumour Biol. J. Int. Soci. Oncodev. Biol. Med.* **2015**, *36*, 6973–6983. [CrossRef] [PubMed]
61. Li, C.; Chen, Y.; Chen, X.; Wei, Q.; Cao, B.; Shang, H. Downregulation of microRNA-193b-3p promotes autophagy and cell survival by targeting TSC1/mTOR signaling in NSC-34 cells. *Front. Mol. Neurosci.* **2017**, *10*, 160. [CrossRef] [PubMed]
62. Denaro, M.; Ugolini, C.; Poma, A.M.; Borrelli, N.; Materazzi, G.; Piaggi, P.; Chiarugi, M.; Miccoli, P.; Vitti, P.; Basolo, F. Differences in miRNA expression profiles between wild-type and mutated NIFTPs. *Endocr.-Relat. Cancer* **2017**, *24*, 543–553. [CrossRef] [PubMed]

Article

Cadmium Exposure Disrupts Periodontal Bone in Experimental Animals: Implications for Periodontal Disease in Humans

Andrew W. Browar [1],*, Emily B. Koufos [1], Yifan Wei [1], Landon L. Leavitt [1], Walter C. Prozialeck [2] and Joshua R. Edwards [2]

[1] College of Dental Medicine, Midwestern University, Illinois, 555 W. 31st St., Science Hall, 211-J, Downers Grove, IL 60515, USA; ekoufos39@midwestern.edu (E.B.K.); ywei89@midwestern.edu (Y.W.); lleavitt83@midwestern.edu (L.L.L.)
[2] Department of Pharmacology, Midwestern University, Downers Grove, IL 60515, USA; wprozi@midwestern.edu (W.C.P.); jedwar@midwestern.edu (J.R.E.)
*   Correspondence: abrowa@midwestern.edu; Tel: +1-630-515-6264

Received: 26 April 2018; Accepted: 11 June 2018; Published: 13 June 2018

Abstract: Cadmium (Cd) is an environmental contaminant that damages the kidney, the liver, and bones. Some epidemiological studies showed associations between Cd exposure and periodontal disease. The purpose of this study was to examine the relationship between Cd exposure and periodontal disease in experimental animals. Male Sprague/Dawley rats were given daily subcutaneous injections of Cd (0.6 mg/kg/day) for up to 12 weeks. The animals were euthanized, and their mandibles and maxillae were evaluated for levels of periodontal bone by measuring the distance from the cementoenamel junction (CEJ) to the alveolar bone crest (ABC) of the molar roots. After 12 weeks of Cd exposure in animals, there was a significantly greater distance between the CEJ and ABC in the palatal aspect of the maxillary molars and the lingual aspect of the mandibular molars when compared with controls ($p < 0.0001$). This study shows that Cd has significant, time-dependent effects on periodontal bone in an animal model of Cd exposure. These findings support the possibility of Cd being a contributing factor to the development of periodontal disease in humans.

Keywords: cadmium; periodontal disease; periodontitis; alveolar bone; osteotoxicity; one health

---

1. Introduction

The World Health Organization recognizes that chronic diseases often share common associations. For example, factors such as diet, lack of physical activity, environmental exposures, and tobacco use are associated with various health problems [1]. Chronic periodontitis is an oral disease that attacks the supporting structures of the teeth (gingiva and jaw bone). Besides affecting oral health, periodontal disease is associated with many other chronic health conditions, including cardiovascular disease and diabetes [2]. Periodontitis is an inflammatory process in response to dental plaque bacteria that activates the innate and adaptive immune responses [3]. It affects approximately 47% of the United States (US) population over 30 years old [4].

The use of tobacco products is a major factor in oral disease, with a dose-related and cumulative relationship with the severity of periodontitis [5]. Smoking was also shown to create a dysbiosis of the oral bacterial flora, favoring more pathogenic bacteria [6].

Of the many toxic components that make up tobacco smoke, the metal cadmium (Cd) is notable in that it is a group 1 carcinogen with toxic effects in lung, liver, testicular, kidney, and bone tissues [7]. Cd is a universally present and naturally occurring environmental contaminant found in a wide variety of common types of food, such as spinach, sunflower seeds, beef liver, and peanuts [8,9].

Diet is a major source of Cd exposure, especially for people living in areas with high levels of Cd contamination [10]. In vivo, Cd exhibits complex toxicokinetics, complicating efforts to understand the mechanisms of Cd toxicity [11,12]. Cd can be absorbed from the lung or the gastrointestinal (GI) tract, before being quickly distributed to the liver where it becomes sequestered upon binding to metallothionein. However, as hepatocytes die from either Cd-induced injury, or general cell turnover, the Cd can redistribute to other tissues, especially the kidney, but also to pancreas and bone. As a result of its tendency to be sequestered in tissues, the half-life for the elimination of Cd from the body is estimated to be as high as 30 years [13]. It is also important to note that blood levels of Cd are usually only elevated during acute, relatively high-level exposure. As Cd accumulates in tissues, blood levels tend to fall. Little or no cadmium is excreted in the urine until the epithelial cells of the proximal tubule are injured by Cd [11,12]. The kidney is considered the primary target organ of Cd toxicity, with concentrations reaching the highest levels in the renal cortex over time. While Cd is known to cause hypercalciuria via renal injury, and lead to osteoporosis via indirect means, Cd also has direct osteotoxic effects on bone tissue, resulting in enhanced bone resorption [14,15]. Biomarkers for bone formation and resorption, such as serum osteocalcin and urinary cross-linked N-telopeptide of type I collagen, were significantly correlated with Cd exposure in a population in Thailand with high dietary Cd intake [16]. Cd having direct and indirect effects on bone formation is significant to the current study because women with osteoporosis are more likely to exhibit periodontal bone loss [17], a hallmark of periodontitis. Cd exposure is also associated with diabetes mellitus and altered metabolic hormone homeostasis [9].

Blood and urinary Cd levels are associated with periodontal disease in the US [18] and South Korea [19,20]. Actual tooth Cd content was higher in a group of patients with periodontal disease [21]. However, other studies did not find significant associations between blood Cd levels and periodontal disease in South Korea [22] or Poland [23]. In addition, the statistical analysis and methodology used by Arora et al. [18] to conclude that a link exists between Cd and periodontal disease were called into question [24].

Because the literature shows contradictory results as to whether Cd exposure is associated with periodontal disease, the goal of the current study was to determine if Cd affects periodontal alveolar bone in a well-established animal model of chronic Cd exposure in rats.

2. Materials and Methods

2.1. Animal Studies

The jaw samples used in the presented study were harvested from rats that were treated with Cd in a series of studies of renal toxicity and urinary biomarkers [25]. All animal studies were conducted in compliance with the United Sates National Institutes of Health (NIH) Guide for the Care and Use of Laboratory Animals (National Research Council of National Academies 2011), and were approved by the Institutional Animal Care and Use Committee of Midwestern University. Adult male Sprague/Dawley rats weighing 250–300 g (Envigo, Indianapolis, IN, USA) were housed socially (two rats per plastic cage) on a 12 h/12 h light/dark cycle. Animals in the Cd treatment group (N = 5–10) received daily (Monday–Friday) subcutaneous injections of $CdCl_2$ at a dose of 0.6 mg (5.36 μmol)/kg in 0.24–0.35 mL isotonic saline for up to 12 weeks. Control group animals (N = 5–10) received daily injections of the saline vehicle only. Animals were euthanized at 6, 9, and 12 weeks, and the tissues were harvested.

2.2. Collection and Preparation of Jaw Samples

Jaws were harvested after animals were anesthetized with an intraperitoneal injection of ketamine/xylazine (67/7 mg/kg) and the kidneys removed. The carcasses were decapitated, and then, the jaws were harvested. Dental scissors (HuFriedy SCGCP) were used to cut through the mandibular symphysis along the floor of the oral cavity lateral to the base of the tongue, and then, through the ramus of the mandible. Buccal musculature and soft tissue were dissected by cutting through the fornix of the buccal vestibule anteriorly. A similar dissection was carried out on the opposite side. The maxillae were harvested by cutting through the palate posterior to the maxillary molars, continuing anteriorly through the fornix of the maxillary vestibule, cutting through the zygomatic process, and across the pre-maxilla through the maxillary incisor teeth. The maxillae were then separated into halves by cutting through the mid-palatal suture. The jaw samples were then placed in labeled containers, and stored in a $-80\ ^{\circ}$C freezer.

Frozen jaw samples (N = 5–10 per treatment group at each time point) were brought to room temperature, defleshed by boiling in water for 7 min to 10 min, manually debrided of soft tissue with periodontal instruments, and then soaked overnight in 5% sodium hypochlorite. The following day, the samples were again carefully debrided of any remaining soft tissue, rinsed, and placed in 3% hydrogen peroxide overnight. Afterward, they were again cleaned, rinsed, and then stained with 1% methylene blue for 1 min, before being rinsed and dried to demarcate the cementoenamel junction (CEJ).

2.3. Morphometric Analysis of Periodontal Bone Levels

Rat jaw segments were affixed to glass microscope slides using soft wax, and viewed using a Nikon E400 microscope with a $2\times$ objective lens, and digital images were captured using an Evolution MP digital air-cooled color camera with the Image Pro Plus image acquisition software (Version 6.1, MediaCybernetics, Rockville, MD, USA, 2006).

To quantify periodontal bone levels, Image Pro Plus image analysis software was utilized. The distance from the CEJ to the alveolar bone crest (ABC) was measured along the main body of each root of each molar (Figure 1). For each molar segment, three values for each first molar root were recorded, and two for each second and third molar (total of seven for each segment). Measurements were taken from the right and left maxillary buccal, maxillary palatal, and mandibular lingual molar segments (six segments per animal).

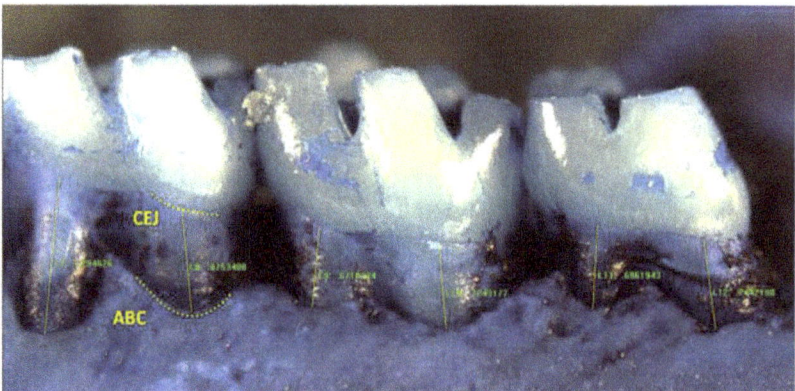

Figure 1. Representative image of a mandibular right lingual molar segment showing measurements from the cementoenamel junction (CEJ) to the alveolar bone crest (ABC) in a rat molar sample. (Mesial measure of first molar not shown).

2.4. Statistical Analysis

Data were analyzed using the Graph Pad Prism statistical program (Version 6.1, La Jolla, CA, USA, 2006). Mean values of the distance from the CEJ to the ABC (mm) for each molar root were evaluated for the saline control versus the Cd-treated animals at the six-week and 12-week time points, using a two-way analysis of variance (ANOVA). If significant differences were detected, a post-hoc Tukey's test was then used to compare values from the time-matched control with the Cd-treated animals. Furthermore, mean measurements from the pooled maxillary buccal, maxillary palatal, and mandibular lingual measurements were also compared. For all analyses, $p \leq 0.05$ was considered as statistically significant.

3. Results

Cd was associated with a time-dependent decrease in periodontal bone levels in an animal model of long-term Cd exposure. The animals tolerated the daily subcutaneous injections of Cd very well, and no animal died or was removed from the study early due to an excessive loss (>20%) of body weight.

Twelve-week samples included two cohorts of experimental and control animals (N = 6 + 4) whereas the six-week sample included only one cohort (N = 5), accounting for the difference in sample size (Table 1).

Mandibular buccal values were not measured because of their proximity to the external oblique ridge and the ascending ramus to the alveolar bone. Nine-week samples were not included because they were not readable. A protracted time (>9 months) in freezer storage prior to processing resulted in the teeth being loose in their alveolar housing, and measurements were not reliable (Table 1).

After 12 weeks of Cd exposure to the experimental animals, there was a significantly greater distance between the cementoenamel junction (CEJ) and the alveolar bone crest (signifying poorer periodontal bone levels) at the palatal aspect of the maxillary molars and the lingual aspect of the mandibular molars when compared with saline-treated control animals ($p < 0.0001$) (Figure 2 and Table 2). A time-dependent change was shown with the maxillary palatal aspect from a comparison of the six-week and 12-week experimental groups ($p < 0.0001$). In the mandibular lingual aspect comparison of the six-week and 12-week experimental groups, the difference was nearly significant ($p = 0.053$).

Table 1. Animals examined at each time point.

Treatment	Week 6	Week 9	Week 12a	Week 12b
N control	5	5 *	6	4 **
N experimental	5	5 *	6	4 **

* Due to a processing error these samples were not analyzed. ** Cohort contained six rats to start. Two were harvested for a histologic study.

Table 2. Mean measurements from pooled maxillary buccal, maxillary palatal, and mandibular lingual values.

Segment	Treatment	6-Week Mean (mm) N = 5 Per Group	SD	12-Week Mean (mm) N = 10 Per Group	SD
Maxilla/Buccal	Control	0.404	±0.134	0.398	±0.151
	Cd-treated	0.431	±0.127	0.441	±0.168
Maxilla/Palatal	Control	0.531	±0.171	0.527	±0.180
	Cd-treated	0.558	±0.216	0.645 * $p < 0.0001$, # $p < 0.0001$	±0.235
Mandible/Lingual	Control	0.820	±0.273	0.785	±0.269
	Cd-treated	0.803	±0.316	0.858 * $p < 0.0001$, # $p = 0.053$	±0.290

All values the distance in mm from the cementoenamel junction (CEJ) to the alveolar bone crest (ABC). * The *p*-value comparing the same week-matched control with the Cd-treated sample. # The *p*-value comparing the six-week sample with the 12-week sample of the same treatment.

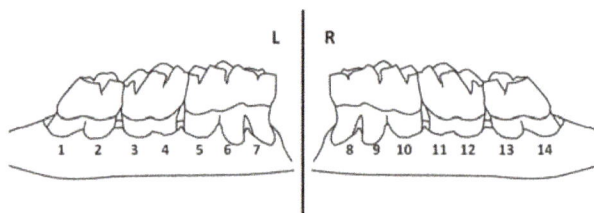

Representative drawing of mandibular lingual segments, showing the molar root identifiers.

CEJ-to-ABC distance in buccal maxilla in the six-week control versus Cd-treated rats (N = 5)

CEJ-to-ABC distance in buccal maxilla in the 12-week control versus Cd-treated rats (N = 10)

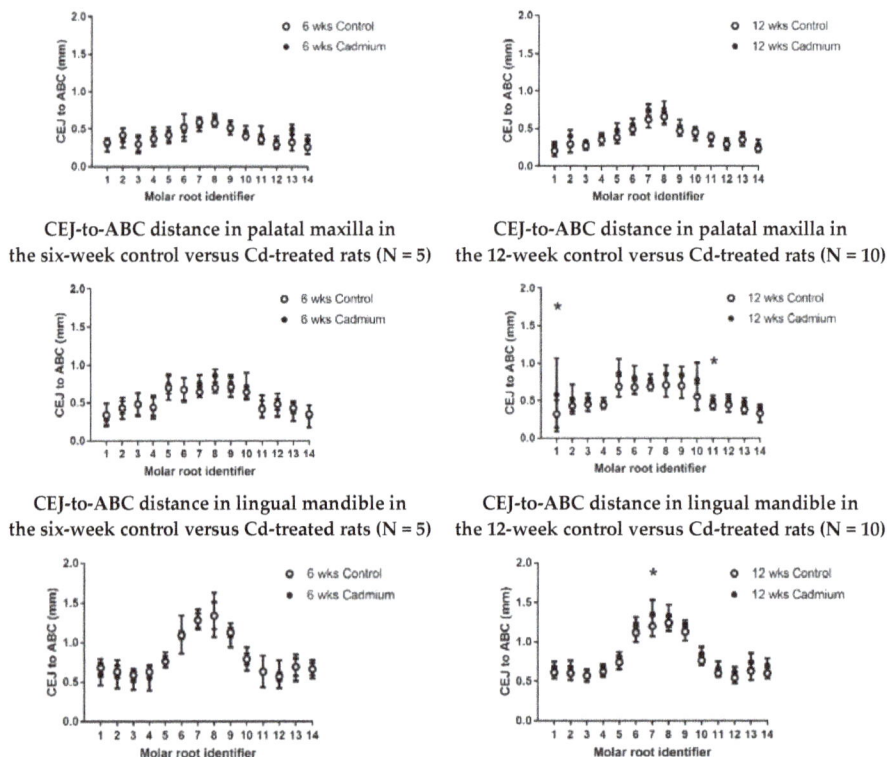

CEJ-to-ABC distance in palatal maxilla in the six-week control versus Cd-treated rats (N = 5)

CEJ-to-ABC distance in palatal maxilla in the 12-week control versus Cd-treated rats (N = 10)

CEJ-to-ABC distance in lingual mandible in the six-week control versus Cd-treated rats (N = 5)

CEJ-to-ABC distance in lingual mandible in the 12-week control versus Cd-treated rats (N = 10)

Figure 2. Graphs comparing the CEJ-to-ABC distances in Cd-treated (0.6 mg/kg/day) animals and saline-treated control animals for individual molar roots. An asterisk (*) indicates significant differences between matching control values, as determined by two-way analysis of variance (ANOVA), followed by Tukey's post-hoc multiple comparison tests; data are given as mean ± SD; N = 5–10.

4. Discussion

The literature shows a number of studies relating the prevalence of periodontal disease to an exposure to environmental Cd [18–20,22,23], with varying methodologies and conclusions. Arora [18] used a partial-mouth method detailed in the Third National Health and Nutrition Examination Survey (NHANES III) [26]. Han [19], Won [20], and Kim [22] used a partial-mouth method from the Community Periodontal Index of Treatment Needs (CPITN) [27]. Herman [23] utilized the CPITN method; however, it was not specified whether a full-mouth or partial-mouth method was used, and their threshold for periodontitis was not identified. Furthermore, the partial-mouth methodology was shown to underestimate the prevalence of disease [28,29], and this potentially affected their outcomes in an unknown way.

This is the first published study examining the effect of Cd on periodontal changes in a controlled animal study. Previously, copper (copper sulfate) exposure in drinking water was investigated in experimental hamsters. That study showed that dietary copper improved periodontal bone levels, particularly in males [30].

The results presented in this study showed that animals given daily doses of Cd (0.6 mg/kg/day) for up to 12 weeks had significant, time-dependent changes of periodontal bone levels, as measured by the CEJ-to-ABC distance. This particular dosing protocol was chosen as it was widely used in studies examining the effects of cadmium on organs such as the kidney, the liver, the pancreas, and bones [12,25,31,32]. This protocol can be used to reproducibly induce various levels of cadmium toxicity, from mild levels at six weeks to severe levels at 12 weeks. In addition, the patterns of Cd distribution and toxicity with this model were comparable to those with chronic oral exposure [11]. Most importantly, since this protocol was used extensively, and is widely accepted as a standard in the Cd field [33–36], it allows for the comparison and interpretation of results across different studies.

Several direct and indirect mechanisms may be responsible for the apparent effects of Cd on periodontal bone levels. Cd exerts direct osteotoxic effects on bone tissue, as well as having indirect effects on bone metabolism and changes in blood calcium regulatory hormones, including parathyroid hormone levels in humans [15]. Numerous studies in the experimental animal and in vitro literature show that Cd in very low doses has direct osteotoxic effects, causing osteoblasts and osteoblast precursor cells to undergo degeneration, while increasing osteoclast formation or activity [14]. In a tissue culture of mouse calvaria osteoblasts, exposure to Cd was found to stimulate a release of calcium from the tissue culture [37,38]. In an earlier study, calcium release from a similar culture was found to be mediated by the production of prostaglandin E2 [39], a potent cytokine for bone resorption. In a study of a peripheral blood mononuclear cell (PBMC) culture, low doses of Cd stimulated the expression of messenger RNA (mRNA) for interleukin 1 (IL-1) and tumor necrosis factor alpha (TNFα), and also of IL-6 at higher doses [40,41], all of which are inflammatory mediators related to periodontitis and periodontal bone loss. In a subsequent study of PBMC cultures and Cd exposure, bacterial antigens of *Salmonella enteritidis* had an immune-modulatory effect on inflammatory mediator expression [42], suggesting that a bacterial challenge coupled with Cd exposure will differentially affect an immunological response to either challenge. These same pro-inflammatory mediators are also associated with cardiovascular disease and diabetes [2].

Another potential mechanism of Cd-induced bone changes may result from Cd-induced renal injury or dysfunction, directly affecting blood calcium levels and subsequent bone metabolism [43]. The animal model of Cd exposure used in this study (0.6 mg/kg/day) was developed to examine the nephrotoxic effects of Cd, with sensitive urinary biomarkers of renal injury, such as kidney injury molecule-1 (Kim-1) and cystatin C, beginning to appear around the fourth week, followed by overt and progressive renal dysfunction starting around the ninth week [25]. It is interesting to note that the change in periodontal bone levels reported here became statistically significant after the time point when overt renal toxicity would be apparent (e.g., after the ninth week). This indicates that some of the Cd effects on the periodontium may be due to indirect effects on whole-body calcium levels due to renal dysfunction.

There are specific aspects of Cd toxicodynamics that need to be taken into consideration when examining the studies showing [18–20] or not showing [22,23] associations between measures of Cd exposure (e.g., blood or urinary Cd) and periodontal disease, namely the source of the sample to determine Cd exposure. Cd is a cumulative toxin with blood or urine levels typically reflecting recent or short-term exposures. This would suggest that the Cd effects on the periodontium occur over time. Elevated urinary Cd levels may also be the result of proximal tubule epithelial cell death, and are not necessarily due to recent exposures, especially in older individuals [12].

When comparing experimental animal models of Cd poisoning and human exposures, it is also important to consider the sources of Cd exposure. Some authors convincingly argued that an overlooked source of human Cd exposure, as related to periodontal disease, is the use of intraoral dental alloys [44]. Human teeth were shown to be a site of Cd accumulation, and presumably, a source of exposure to adjacent alveolar bone. In a study of both smokers and non-smokers with periodontal disease, tooth Cd levels were ten-fold or higher when compared with teeth from patients who did not have periodontal disease [21]. Therefore, the possibility exists that Cd is absorbed into teeth from dental restorations or tobacco smoke, and may have direct effects on neighboring periodontal bone.

This study is currently limited to male Sprague/Dawley rats. Morphometric studies investigating periodontal bone changes with other experimental studies (not involving Cd) showed a gender difference in responses [30,45], and Cd also showed gender-specific changes in bone responses in humans [16]. Future studies should examine the effects of Cd on periodontal bone loss in both male and female animals.

There was a significantly greater distance between the CEJ and the ABC (signifying poorer periodontal bone levels) when comparing the 12-week Cd-exposed animals with the saline-treated controls ($p < 0.0001$) at the palatal aspect of the maxillary molars, and the lingual aspect of the mandibular molars. This was not observed after six weeks (Table 2). It is worth noting that bone levels for the 12-week maxillary buccal Cd-treated animals did not reach a level of statistical significance when compared with controls. The anatomy of the periodontal bone at the alveolar crest is somewhat thicker on the buccal aspect of the maxilla than in the other sites, and may explain the results reported here. This difference between buccal and palatal or lingual changes was reported in other morphometric experimental studies (not involving Cd) in rats and mice [45–47]. The pattern of bone changes in the six-week and 12-week experimental groups were remarkably uniform, with a consistent level of bone crest for each tooth and/or root, and no significant changes in interproximal bone levels were observed. (e.g., see Figure 2 diagram between root identifiers 2–3, 4–5, 10–11, and 12–13). When comparing the maxillary palatal six-week to 12-week experimental groups, there was a significant difference ($p < 0.0001$). When comparing the mandibular lingual six-week to 12-week experimental groups, the difference was nearly significant ($p = 0.053$). As we replicate the study and increase the N values, we anticipate that the data will show more significant differences (Table 2).

While the results in this study show the effects of experimental Cd exposure on periodontal bone levels, the findings do not address the etiology of Cd-induced periodontal bone changes. Further study is needed to identify the mechanism(s) responsible for Cd-mediated periodontal bone loss, as well as the effects age, gender, systemic disease (e.g., diabetes and cardiovascular disease), and local factors (e.g., bacterial challenge, smoking, or toxicity from dental restorations) have on the Cd–periodontium interaction.

Author Contributions: A.W.B., J.R.E., and W.C.P. conceived and designed the experiments, interpreted the results, assisted with the animal study, and wrote the paper. E.B.K., Y.W., and L.L.L. prepared harvested tissues for image and Cd analysis, and performed the measurements and prepared the tables for analysis.

Funding: The research presented in this manuscript was supported by the College of Dental Medicine—Illinois intramural research funds at Midwestern University and by the National Institutes of Health grant number R15 ES028443 to J.R.E.

Acknowledgments: Thank you Victoria Sears, Laura Phelps, and Taisa Szeremeta-Browar, DDS for assistance with preparing the manuscript, figures, editing, and the use of EndNote.

Conflicts of Interest: The authors declare no conflicts of interest.

References

1. World Health Organization. Facing the Facts #1, Chronic Diseases and Their Common Risk Factors. Available online: http://www.who.int/chp/chronic_disease_report/media/Factsheet1.pdf (accessed on 26 January 2018).
2. Burt, B. Position paper: Epidemiology of periodontal diseases. *J. Periodontol.* **2005**, *76*, 1406–1419. [PubMed]
3. Cekici, A.; Kantarci, A.; Hasturk, H.; Van Dyke, T.E. Inflammatory and immune pathways in the pathogenesis of periodontal disease. *Periodontology 2000* **2014**, *64*, 57–80. [CrossRef] [PubMed]
4. Eke, P.I.; Dye, B.A.; Wei, L.; Thornton-Evans, G.O.; Genco, R.J. CDC Periodontal Disease Surveillance workgroup: James Beck, Gordon Douglass, Roy Page, 2012. Prevalence of periodontitis in adults in the United States: 2009 and 2010. *J. Dent. Res.* **2012**, *91*, 914–920. [CrossRef] [PubMed]
5. Tomar, S.L.; Asma, S. Smoking-attributable periodontitis in the United States: Findings from NHANES III. *J. Periodontol.* **2000**, *71*, 743–751. [CrossRef] [PubMed]
6. Borojevic, T. Smoking and periodontal disease. *Mater. Sociomed.* **2012**, *24*, 274–276. [CrossRef] [PubMed]
7. Jarup, L.; Berglund, M.; Elinder, C.G.; Nordberg, G.; Vahter, M. Health effects of cadmium exposure—A review of the literature and a risk estimate. *Scand. J. Work Environ. Health* **1998**, *24* (Suppl. 1), 1–51. [PubMed]
8. Paschal, D.C.; Burt, V.; Caudill, S.P.; Gunter, E.W.; Pirkle, J.L.; Sampson, E.J.; Miller, D.T.; Jackson, R.J. Exposure of the U.S. population aged 6 years and older to cadmium: 1988–1994. *Arch. Environ. Contam. Toxicol.* **2000**, *38*, 377–383. [CrossRef] [PubMed]
9. Edwards, J.; Ackerman, C. A review of diabetes mellitus and exposure to the environmental toxicant cadmium with an emphasis on likely mechanisms of action. *Curr. Diabetes Rev.* **2016**, *12*, 252–258. [CrossRef] [PubMed]
10. Satarug, S.; Swaddiwudhipong, W.; Ruangyuttikarn, W.; Nishijo, M.; Ruiz, P. Modeling cadmium exposures in low- and high-exposure areas in Thailand. *Environ. Health Perspect.* **2013**, *121*, 531–536. [CrossRef] [PubMed]
11. Prozialeck, W.C.; Edwards, J.R. Early biomarkers of cadmium exposure and nephrotoxicity. *Biometals* **2010**, *23*, 793–809. [CrossRef] [PubMed]
12. Prozialeck, W.C.; Edwards, J.R. Mechanisms of cadmium-induced proximal tubule injury: New insights with implications for biomonitoring and therapeutic interventions. *J. Pharmacol. Exp. Ther.* **2012**, *343*, 2–12. [CrossRef] [PubMed]
13. Friberg, L. Cadmium and the kidney. *Environ. Health Perspect.* **1984**, *54*, 1–11. [CrossRef] [PubMed]
14. Bhattacharyya, M.H. Cadmium osteotoxicity in experimental animals: Mechanisms and relationship to human exposures. *Toxicol. Appl. Pharmacol.* **2009**, *238*, 258–265. [CrossRef] [PubMed]
15. Schutte, R.; Nawrot, T.S.; Richart, T.; Thijs, L.; Vanderschueren, D.; Kuznetsova, T.; Van Hecke, E.; Roels, H.A.; Staessen, J.A. Bone resorption and environmental exposure to cadmium in women: A population study. *Environ. Health Perspect.* **2008**, *116*, 777–783. [CrossRef] [PubMed]
16. Nishijo, M.; Nambunmee, K.; Suvagandha, D.; Swaddiwudhipong, W.; Ruangyuttikarn, W.; Nishino, Y. Gender-specific impact of cadmium exposure on bone metabolism in older people living in a cadmium-polluted area in Thailand. *Int. J. Environ. Res. Public Health* **2017**, *14*, 401. [CrossRef] [PubMed]
17. Penoni, D.C.; Fidalgo, T.K.; Torres, S.R.; Varela, V.M.; Masterson, D.; Leao, A.T.; Maia, L.C. Bone density and clinical periodontal attachment in postmenopausal women: A systematic review and meta-analysis. *J. Dent. Res.* **2017**, *96*, 261–269. [CrossRef] [PubMed]
18. Arora, M.; Weuve, J.; Schwartz, J.; Wright, R.O. Association of environmental cadmium exposure with periodontal disease in U.S. adults. *Environ. Health Perspect.* **2009**, *117*, 739–744. [CrossRef] [PubMed]
19. Han, D.H.; Lee, H.J.; Lim, S. Smoking induced heavy metals and periodontitis: Findings from the Korean National Health and Nutrition Examination Surveys 2008–2010. *J. Clin. Periodontol.* **2013**, *40*, 850–858. [CrossRef] [PubMed]
20. Won, Y.S.; Kim, J.H.; Kim, Y.S.; Bae, K.H. Association of internal exposure of cadmium and lead with periodontal disease: A study of the fourth Korean National Health and Nutrition Examination Survey. *J. Clin. Periodontol.* **2013**, *40*, 118–124. [CrossRef] [PubMed]

21. Alhasmi, A.M.; Gondal, M.A.; Nasr, M.M.; Shafik, S.; Habibullah, Y.B. Detection of toxic elements using laser-induced breakdown spectroscopy in smokers' and nonsmokers' teeth and investigation of periodontal parameters. *Appl. Opt.* **2015**, *54*, 7342–7349. [CrossRef] [PubMed]
22. Kim, Y.; Lee, B.K. Association between blood lead and mercury levels and periodontitis in the Korean general population: Analysis of the 2008–2009 Korean national health and nutrition examination survey data. *Int. Arch. Occup. Environ. Health* **2013**, *86*, 607–613. [CrossRef] [PubMed]
23. Herman, M.; Golasik, M.; Piekoszewski, W.; Walas, S.; Napierala, M.; Wyganowska-Swiatkowska, M.; Kurhanska-Flisykowska, A.; Wozniak, A.; Florek, E. Essential and toxic metals in oral fluid—A potential role in the diagnosis of periodontal diseases. *Biol. Trace Elem. Res.* **2016**, *173*, 275–282. [CrossRef] [PubMed]
24. Dye, B.A.; Dillon, C.F. Elevated cadmium exposure may be associated with periodontal bone loss. *J. Evid. Based Dent. Pract.* **2010**, *10*, 109–111. [CrossRef] [PubMed]
25. Prozialeck, W.C.; VanDreel, A.; Ackerman, C.D.; Stock, I.; Papaeliou, A.; Yasmine, C.; Wilson, K.; Lamar, P.C.; Sears, V.L.; Gasiorowski, J.Z.; et al. Evaluation of cystatin C as an early biomarker of cadmium nephrotoxicity in the rat. *Biometals* **2016**, *29*, 131–146. [CrossRef] [PubMed]
26. Albandar, J.M.; Brunelle, J.A.; Kingman, A. Destructive periodontal disease in adults 30 years of age and older in the United States, 1988–1994. *J. Periodontol.* **1999**, *70*, 13–29. [CrossRef] [PubMed]
27. Ainamo, J.; Barmes, D.; Beagrie, G.; Cutress, T.; Martin, J.; Sardo-Infirri, J. Development of the World Health Organization (WHO) Community Periodontal Index of Treatment needs (CPITN). *Int. Dent. J.* **1982**, *32*, 281–291. [PubMed]
28. Eke, P.I.; Thornton-Evans, G.O.; Wei, L.; Borgnakke, W.S.; Dye, B.A. Accuracy of NHANES periodontal examination protocols. *J. Dent. Res.* **2010**, *89*, 1208–1213. [CrossRef] [PubMed]
29. Miller, N.A.; Benamghar, L.; Roland, E.; Martin, G.; Penaud, J. An analysis of the community periodontal index of treatment needs. Studies on adults in France. III—Partial examinations versus full-mouth examinations. *Community Dent. Health* **1990**, *7*, 249–253. [PubMed]
30. Costich, E.R. A Quantitative Evaluation of the Effect of Copper on Alveolar Bone Loss in the Syrian Hamster. *J. Periodontol.* **1955**, *26*, 301–305. [CrossRef]
31. Prozialeck, W.C.; Vaidya, V.S.; Liu, J.; Waalkes, M.P.; Edwards, J.R.; Lamar, P.C.; Bernard, A.M.; Dumont, X.; Bonventre, J.V. Kidney injury molecule-1 is an early biomarker of cadmium nephrotoxicity. *Kidney Int.* **2007**, *72*, 985–993. [CrossRef] [PubMed]
32. Prozialeck, W.C.; Edwards, J.R.; Lamar, P.C.; Liu, J.; Vaidya, V.S.; Bonventre, J.V. Expression of kidney injury molecule-1 (Kim-1) in relation to necrosis and apoptosis during the early stages of Cd-induced proximal tubule injury. *Toxicol. Appl. Pharmacol.* **2009**, *238*, 306–314. [CrossRef] [PubMed]
33. Aoyagi, T.; Hayakawa, K.; Miyaji, K.; Ishikawa, H.; Hata, M. Cadmium nephrotoxicity and evacuation from the body in a rat modeled subchronic intoxication. *Int. J. Urol.* **2003**, *10*, 332–338. [CrossRef] [PubMed]
34. Dudley, R.E.; Gammal, L.M.; Klaassen, C.D. Cadmium-induced hepatic and renal injury in chronically exposed rats: Likely role of hepatic cadmium-metallothionein in nephrotoxicity. *Toxicol. Appl. Pharmacol.* **1985**, *77*, 414–426. [CrossRef]
35. Goyer, R.A.; Miller, C.R.; Zhu, S.Y.; Victery, W. Non-metallothionein-bound cadmium in the pathogenesis of cadmium nephrotoxicity in the rat. *Toxicol. Appl. Pharmacol.* **1989**, *101*, 232–244. [CrossRef]
36. Shaikh, Z.A.; Northup, J.B.; Vestergaard, P. Dependence of cadmium-metallothionein nephrotoxicity on glutathione. *J. Toxicol. Environ. Health A* **1999**, *57*, 211–222. [PubMed]
37. Carlsson, L.; Lundholm, C.E. Characterisation of the effects of cadmium on the release of calcium and on the activity of some enzymes from neonatal mouse calvaria in culture. *Comp. Biochem. Physiol. C Pharmacol. Toxicol. Endocrinol.* **1996**, *115*, 251–256. [CrossRef]
38. Romare, A.; Lundholm, C.E. Cadmium-induced calcium release and prostaglandin E2 production in neonatal mouse calvaria are dependent on COX-2 induction and protein kinase C activation. *Arch. Toxicol.* **1999**, *73*, 223–228. [CrossRef] [PubMed]
39. Suzuki, Y.; Morita, I.; Yamane, Y.; Murota, S. Cadmium stimulates prostaglandin E2 production and bone resorption in cultured fetal mouse calvaria. *Biochem. Biophys. Res. Commun.* **1989**, *158*, 508–513. [CrossRef]
40. Marth, E.; Barth, S.; Jelovcan, S. Influence of cadmium on the immune system. Description of stimulating reactions. *Cent. Eur. J. Public Health* **2000**, *8*, 40–44. [PubMed]
41. Marth, E.; Jelovcan, S.; Kleinhappl, B.; Gutschi, A.; Barth, S. The effect of heavy metals on the immune system at low concentrations. *Int. J. Occup. Med. Environ. Health* **2001**, *14*, 375–386. [PubMed]

42. Hemdan, N.Y.; Emmrich, F.; Sack, U.; Wichmann, G.; Lehmann, J.; Adham, K.; Lehmann, I. The in vitro immune modulation by cadmium depends on the way of cell activation. *Toxicology* **2006**, *222*, 37–45. [CrossRef] [PubMed]

43. Horiguchi, H.; Oguma, E.; Sasaki, S.; Miyamoto, K.; Ikeda, Y.; Machida, M.; Kayama, F. Environmental exposure to cadmium at a level insufficient to induce renal tubular dysfunction does not affect bone density among female Japanese farmers. *Environ. Res.* **2005**, *97*, 83–92. [CrossRef] [PubMed]

44. Guzzi, G.; Pigatto, P.D.; Ronchi, A. Periodontal disease and environmental cadmium exposure. *Environ. Health Perspect.* **2009**, *117*, A535–A536. [CrossRef] [PubMed]

45. Rivaldo, E.G.; Padilha, D.P.; Hugo, F.N. Alveolar bone loss and aging: A model for the study in mice. *J. Periodontol.* **2005**, *76*, 1966–1971. [CrossRef] [PubMed]

46. Abe, T.; Hajishengallis, G. Optimization of the ligature-induced periodontitis model in mice. *J. Immunol. Methods* **2013**, *394*, 49–54. [CrossRef] [PubMed]

47. Crawford, J.M.; Taubman, M.A.; Smith, D.J. The natural history of periodontal bone loss in germfree and gnotobiotic rats infected with periodontopathic microorganisms. *J. Periodontal Res.* **1978**, *13*, 316–325. [CrossRef] [PubMed]

Article

Mitochondrial Morphology and Function of the Pancreatic β-Cells INS-1 Model upon Chronic Exposure to Sub-Lethal Cadmium Doses

Adeline Jacquet [1], Cécile Cottet-Rousselle [1], Josiane Arnaud [1,2], Kevin Julien Saint Amand [1], Raoua Ben Messaoud [1], Marine Lénon [1], Christine Demeilliers [1] and Jean-Marc Moulis [1,3,*]

[1] Laboratory of Fundamental and Applied Bioenergetics (LBFA), Inserm, Universite Grenoble Alpes, 38000 Grenoble, France; adeline.jacquet1@orange.fr (A.J.); cecile.cottet@univ-grenoble-alpes.fr (C.C.-R.); jarnaud@chu-grenoble.fr (J.A.); jsakevin@gmail.com (K.J.S.A.); raouabenmessaoud@yahoo.com (R.B.M.); marine.lenon@hotmail.fr (M.L.); christine.demeilliers@univ-grenoble-alpes.fr (C.D.)

[2] Biochemistry, Molecular Biology and Environmental Toxicology (SB2TE), Grenoble University Hospital, CS 10217, 38043 Grenoble, France

[3] CEA-Grenoble, Bioscience and Biotechnology Institute (BIG), 38054 Grenoble, France

* Correspondence: jean-marc.moulis@cea.fr; Tel.: +33-476-635-536

Received: 19 February 2018; Accepted: 20 March 2018; Published: 22 March 2018

Abstract: The impact of chronic cadmium exposure and slow accumulation on the occurrence and development of diabetes is controversial for human populations. Islets of Langerhans play a prominent role in the etiology of the disease, including by their ability to secrete insulin. Conversion of glucose increase into insulin secretion involves mitochondria. A rat model of pancreatic β-cells was exposed to largely sub-lethal levels of cadmium cations applied for the longest possible time. Cadmium entered cells at concentrations far below those inducing cell death and accumulated by factors reaching several hundred folds the basal level. The mitochondria reorganized in response to the challenge by favoring fission as measured by increased circularity at cadmium levels already ten-fold below the median lethal dose. However, the energy charge and respiratory flux devoted to adenosine triphosphate synthesis were only affected at the onset of cellular death. The present data indicate that mitochondria participate in the adaptation of β-cells to even a moderate cadmium burden without losing functionality, but their impairment in the long run may contribute to cellular dysfunction, when viability and β-cells mass are affected as observed in diabetes.

Keywords: mitochondrial network; image analysis; mitochondrial morphology; bioenergetics; sub-lethal exposure; toxicological mechanism; cadmium

---

1. Introduction

As a widespread contaminant found in the environment, the metal cadmium and its toxicity remain the subject of continuous studies with different approaches. Indeed, low background cadmium contamination in soils, usually below 0.5 ppm, is contributed by dispersion from various sources, such as mining and refining of other metals, waste disposal, and the dispersal of phosphate fertilizers. Increased environmental impact occurs at some particular sites, such as smelters and waste incinerators [1]. Cadmium dispersion contaminates crops with cadmium passing through the food chain up to farming animals and human populations. Hence, food is a considerable source of cadmium exposure for humans. Recommendations have been issued and continue to be updated by regulatory bodies to minimize the health effect of cadmium contamination of food and drinking water [2,3]. But a major difficulty of the topic is that biological targets of cadmium are numerous [4], and the mechanisms of action of the Cd^{2+} cation, the only ionic state of biological relevance, can be complex [5,6]. Biochemically, this is mainly due to the electropolarity and the size of Cd^{2+} which enable a range of molecular interactions [6,7].

In animals and at the organ level, kidneys, liver, and the skeleton are well-established sites of deposition with cadmium-induced damage or functional impairment [8]. But other organs may also be affected by cadmium exposure [9] and the functional consequences are often difficult to delineate, particularly at low levels of exposure. A case in point is pancreas. It is a site where cadmium can be found [10], in the endocrine compartment, and even in populations without particularly high exposure [11]. Thus, the causative link between cadmium exposure and prevalence of type II diabetes is difficult to establish for human populations [12–14]. As an alternative to the screening of human populations, animal studies may bring illuminating information forth, but too many of them implement unduly high cadmium doses and inappropriate modes of exposure that do not help clarifying the debate [15].

Yet, for the above recalled biochemical reasons, cadmium may well be interfering with the function of the endocrine pancreas, and with insulin production in particular. The later process is the specialized function of β-cells enclosed in islets of Langerhans in response to circulating glucose concentrations. The general model (e.g., [16]) leads from glucose uptake to secretion of insulin granules with increased glycolysis and adenosine triphosphate (ATP) production, closure of potassium channels at the plasma membrane, calcium uptake, and stimulated exocytosis. Most, if not all, steps of this complex mechanism may be sensitive to the presence of cadmium, but which is/are the most susceptible remains unclear. This questioning is particularly worth considering when interest is focused on low levels of exposure [8,15,17], as is the most prevalent situation for human populations nowadays and for which mechanistic insight is expected to provide means for educated intervention modalities.

Previous work has targeted mitochondria as key organelles converting chemical (glucose) into electrical (membrane depolarization) signals via ATP in β-cells [16]. This segment lies in the upper region of the insulin secretion cellular process, and it is thus expected to have downstream consequences if deficient. Short term (24 h) and relatively high cadmium concentration exposure (above the median lethal dose at more than 5 μM) induced apoptosis of the β-cells rodent model RIN-m5F with dysfunction of mitochondria [18] as could be predicted [19,20]. However, the specific impact of doses of cadmium largely below the onset of cell death on the mitochondria of β-cells has never been reported in details. The present work is an attempt to make up for this gap by exploring the morphological and functional changes occurring to mitochondria of INS-1 cells as a function of the longest possible time of exposure to largely sub-lethal concentrations of the Cd^{2+} cation. We tried to separate cellular events occurring either largely before or upon cell death. It thus appeared that mitochondria do sense moderate levels of cadmium, but cells can cope with such a challenge, and functional consequences are only observed when cells begin to die.

2. Materials and Methods

2.1. Cells and Treatments

INS-1 is a pancreatic β-cell line which was isolated from X-ray induced rat insulinoma [21]. It was obtained from the Department of Genetic Medicine and Development, University of Geneva Medical Center (Switzerland). They were maintained in RPMI 1640 medium containing 2 g/L glucose, supplemented with 10% foetal bovine serum (FBS), antibiotics (1% of penicillin (10,000 U/mL) and streptomycin (10 mg/mL)), 50 μM 2 mercaptoethanol, 1 mM sodium pyruvate and 2 mM Glutamine (complete medium) and grown in a humidified incubator with 5% CO_2 at 37 °C.

A $CdCl_2$ solution (250 μM) in PBS was added to the medium at appropriate volumes to achieve the intended end concentrations after cell adhesion. Seeding was at 1.5×10^5 cells/mL and growth occurred either in complete medium (control) or in complete medium containing the required concentration of $CdCl_2$. Incubation was for 72 or 96 h with regular medium replacement at 37 °C, 5% CO_2.

2.2. Viability Measurements

To assess viability, INS-1 cells were exposed to CdCl$_2$ in the complete culture medium at different concentrations in the [0–20] μM range for 72 h. At the end of the treatment, cells were washed with PBS and brought in suspension by incubation with 0.25% trypsin/EDTA in PBS without calcium and magnesium. Cells were pelleted at 150× *g* for 5 min, the pellet was rinsed with PBS, and suspended at 10^6 cells/mL in 50 mM HEPES, 0.7 M NaCl, 12.5 mM CaCl$_2$, pH 7.4. The suspension was labeled with Fluoprobe 488-annexin V (Interchim) then 1 μg/mL propidium iodide (PI) for 15 min at room temperature in the dark. The stained cells were detected by flow cytometry with a LSR Fortessa™ cell analyzer (Becton Dickinson, Le Pont de Claix, France) using the 488 nm sapphire laser and 532 nm compass laser for Fluoprobe 488 and PI, respectively. The corresponding fluorescence emission was measured with a 525/50 nm and 585/15 nm band-pass filters, respectively. Live cells are not labeled in this assay, whereas preapoptotic ones bind annexin V, necrotic ones accumulate PI, and doubly labeled cells are the dead ones. As an alternative method to the above labeling of cells, viability was also measured with the Cell Titer 96® AQ$_{ueous}$ One Solution Cell Proliferation Assay (Promega, Madison, WI, USA) in 96 well plates until adherence, then cadmium was added at different concentrations as explained above. The number of cells able to reduce the MTS tetrazolium compound was determined by recording the absorbance at 490 nm with a multi-well plate reader (Clariostar, BMG Labtech, Ortenberg, Germany).

2.3. Immunofluorescence Measurements

In immunofluorescence (IF) experiments, INS-1 cells were inoculated at 5000 cells/well on culture slides with detachable culture chambers (Falcon/Corning) until adherence. They were treated with different concentrations of CdCl$_2$ for 96 h as described above. In wells in which mitochondria were labeled without nuclear staining, the cell-permeable fluorescent probe MitoTracker Red CMXRos (ThermoFisher, Illkirch, France) was added at 200 nM for 30 min at 37 °C. Cells were fixed in fresh 4% paraformaldehyde for 10 min at ambient temperature, washed twice with PBS, then cells were permeabilized using 0.2% Triton X-100 in PBS for 15 min, rinsed thrice, and blocked with PBS-Tween (1 mg/mL) BSA 5% (PBS-T BSA) for 1 h at 37 °C. Mitochondria were alternatively labeled with the primary antibody (D6D9 Rabbit mAb, Cell Signaling Technology, Danvers, MA, USA) raised in rabbit against mitochondrial aconitase (the product of the ACO$_2$ gene) as an alternative to MitoTracker staining. The mAb was diluted 200 fold in PBS-T BSA and cells were incubated overnight at 4 °C. The cells were then rinsed thrice with PBS, and the primary antibody was reacted for 4 h at room temperature in the dark with the labeled secondary one (goat anti rabbit secondary antibody Hylite Fluor® 488, Anaspec–Eurogentec, Angers, France) diluted 200 fold in PBS-T BSA. Before the end of the latter incubation, nuclear staining was performed with PI (1 mg/mL) for 20 min at 4 °C. Culture chambers were removed, and slides were mounted and sealed before microscopic observation.

A Leica TCS SP8 inverted laser scanning confocal microscope (Leica Microsystems, Wetzlar, Germany) equipped with a 40× Oil immersion objective was used to collect images. Laser excitation was 488 nm for Hylite Fluor 488, 552 nm for MitoTracker Red CMXRos and PI, with fluorescence emission at 500–550 nm, 575–630 nm, and 605–685 nm, respectively. The Mitotracker probe was used to cross check the images recorded by labeling aconitase: both sets of images qualitatively agreed and, since the latter were of better quality than the former, only wells in which aconitase was detected were analyzed in details. Several fields were recorded for each slide and quantitative analysis with the Image J (imagej.nih.gov) and Volocity (Improvision, Perkin-Elmer, Courtaboeuf, France) computer programs was carried out on all of them as follows. In a first step, tophat filtering was applied to the images recorded with the mitochondrial channel (aconitase fluorescence) in Image J to remove noise and to obtain a precise definition of the mitochondrial morphology. The filtered images were then analyzed with the Volocity software which provides morphological parameters like perimeter, area, skeletal length and diameter for each identified object. Each analyzed Cd-treatment group corresponded to tens of cells, and hundreds or thousands of mitochondrial objects. From these data, the circularity shape factor was calculated, as Equation (1).

$$\text{circularity shape factor} = 4\pi \times \text{mean area}/(\text{mean perimeter})^2 \tag{1}$$

2.4. ATP Measurements

To measure the cellular content in adenosine nucleotides, cells were rinsed with PBS after the cadmium treatment as described above. Cellular perchlorate extracts were prepared with 750 µL of cold 2.5% (*w:v*) HClO$_4$-EDTA 6.25 mM. The recovered mixture was strongly mixed and centrifuged at 12,000× *g* for 5 min at 4 °C. The supernatant was neutralized at pH 7 with MOPS-KOH buffer 0.3 M and centrifuged 2 min at 12,000× *g*. Aliquots (75 µL) of the supernatant were then mixed with 15 µL of HCl 1 M and 35 µL of 28 mM pyrophosphate buffer pH 5.75. Thirty µl of the mixture were analyzed on a Polaris 5 C18-A, Agilent S (250 × 4.6 mm) column equilibrated in pyrophosphate buffer pH 5.75 at 1 mL/min. The retention times of ATP, adenosine diphosphate (ADP) and adenosine monophosphate (AMP) are approximately 6, 7 and 12 min, respectively, as determined by UV absorption at 254 nm.

2.5. Respiration Rates

The oxygen consumption rates of INS-1 cells were measured with a temperature-controlled Hansatech oxygraph equipped with a Clark electrode and monitored with the *oxygraph* software. For measurements with glucose stimulation, harvested cells were suspended in Krebs-Ringer Bicarbonate HEPES buffer (KRBH: NaCl 125 mM, CaCl$_2$ 1 mM, MgSO$_4$ 1.2 mM, KCl 4.74 mM, NaHCO$_3$ 5 mM, BSA 0.1%, pH 7.4) with 2.8 or 16.7 mM of glucose at ca. 4×10^6 cells/mL. These glucose concentrations were chosen at the lower and higher ends of the sigmoid response of β-cells to glucose. Cellular respiration was measured and, when required, 1 µg/mL oligomycin was added. The fraction of O$_2$ consumption used for producing ATP was calculated as Equation (2).

$$\text{fraction of O}_2 \text{ consumption} = (\text{basal} - \text{oligomycin})/\text{basal} \tag{2}$$

2.6. Other Measurements

To estimate the level of oxidative species present inside cells at the end of the exposure period to cadmium, INS-1 cells were grown and treated with CdCl2 as above (Section 2.1) for 96 h, harvested and rinsed with PBS (1 mL/0.5 × 10^6 cells). Dihydroethidium (DHE) at the final concentration of 5 µM was added to the cell suspension which was then incubated 30 min in the dark at 37 °C, and cells were analyzed for fluorescence with a LSR Fortessa™ cell analyzer (Becton Dickinson, Le Pont de Claix, France).

To measure the concentration of cadmium inside cells, the latter were harvested, rinsed and dry pellets were kept at −80 °C before measurements. The latter were carried out by Inductively-coupled plasma-mass spectrometry (ICP-MS) as described in details elsewhere [22]. An aliquot of the cell preparation was lysed and the protein concentration was measured with the bicinchoninic acid method (Uptima–Interchim) to calibrate the results.

2.7. Statistical Analysis

The implemented statistical tests were adjusted to the design of the different experiments, the nature of the measured parameters, and the kind of comparison to be made. The Kruskal–Wallis test or One Way Analysis of Variance, depending on the data distributions, were applied to the parameters tested to vary with the cadmium exposure group. For microscopic image analysis, the Dunn's test was used to compare each group to the one which was not exposed to cadmium. The difference of rank means, the Q test statistic, and the $p < 0.05$ status were recorded.

3. Results

3.1. Effects of Cadmium on INS-1 Cells

3.1.1. Viability of INS-1 Cells upon Long-Term Exposure to Cadmium

Since the aim of the present study is to probe the sensitivity of the INS-1 cell line upon long-term exposure to sub-lethal concentrations of cadmium, these β-cells were kept for the maximal amount of time in culture in the presence of cadmium before analysis. It was observed that a given batch of cells could not sustain growth for more than ca. 4 days before reaching high density for those not exposed or dying for the most exposed ones, i.e., without applying the stress associated with passage in suspension. Hence, 4 days was the time limit set for the following experiments. The analysis of viability after 72 h indicated that the proportion of viable cells was not significantly altered below 2.5 µM (Figure 1). By further keeping cells for an additional day in the presence of cadmium, the decreasing fraction of viable cells as a function of the cadmium concentration was measured by the ability of cells to reduce a tetrazolium compound. The median lethal dose was calculated at 5.0 µM with a standard error of 1.35 in several experiments.

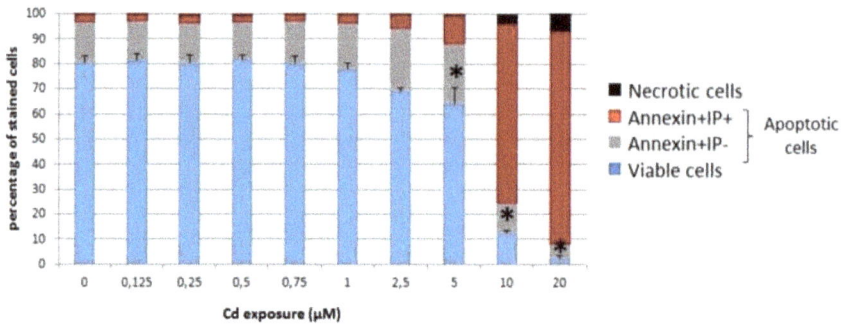

Figure 1. Viability of INS-1 cells in the presence of cadmium. The fractions of viable and death committed cells after 72 h of cadmium exposure were measured by flow cytometry after FluoProbe 488-Annexin V and PI labeling. Only above 2.5 µM was the fraction of viable cells significantly altered. * ANOVA test for live cells $p < 0.05$ vs. 0 µM Cd^{2+} ($n = 4$).

From the above, it was decided that only cadmium concentrations well below the median lethal dose would be worth considering in an effort to mimic chronic low-level exposure. Therefore, further data were obtained by exposing cells to concentrations up to 2 or 2.5 µM, for 96 and 72 h, respectively, these upper points being taken as limits corresponding to the onset of cell death, of less than 15% in each case.

3.1.2. Cadmium Uptake

Despite the sub-lethal cadmium concentrations used in the experimental setup defined in Section 3.1.1, cadmium did enter and accumulated inside cells in a dose-dependent way (Figure 2).

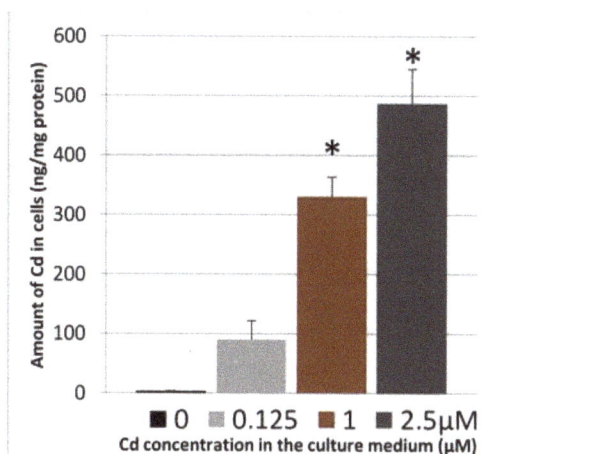

Figure 2. Cadmium uptake by INS-1 cells. The amount of cadmium in washed cells was measured by inductively-coupled plasma-mass spectrometry (ICP-MS) after 72 h of cadmium exposure and normalized to the protein concentration. * ANOVA test, $p < 0.001$ vs. 0 μM Cd^{2+} ($n = 3$).

3.2. Organization of the Mitochondrial Network and Impact of Long-Term Cadmium Exposure

3.2.1. Microscopic Examination of Mitochondria

The visualization of mitochondria by immunofluorescence staining showed that the aspect of the mitochondrial network changed after exposure of cells to doses of cadmium for 96 h (Figure 3). The continuous interconnected reticulum seen for non-exposed cells was gradually transformed into a set of discrete and isolated organelles. The strings of mitochondria disappeared fully at the highest cadmium concentration of 2 μM.

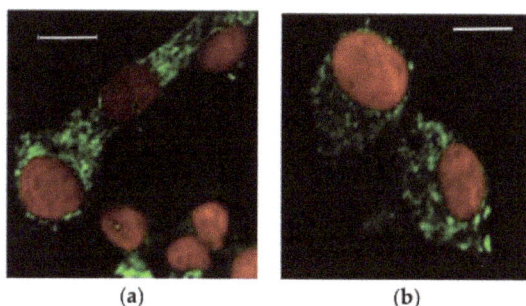

Figure 3. Selected images showing the effect of sub-lethal doses of cadmium on INS-1 cells mitochondria. Mitochondrial morphology was followed by staining a mitochondrial protein of the matrix, aconitase (green). The nucleus was stained by PI (red). INS-1 cells were treated for 96 h with (**a**) 0 and (**b**) 2 μM $CdCl_2$. Scale bars 10 μm.

3.2.2. Quantitative Analysis of the Mitochondrial Network

The microscopic images were treated as described in the Material and Methods section. Two parameters, circularity and skeletal length, were selected for analysis, and the results of comparisons between cadmium treatment groups and control INS-1 cells are shown in Figure 4. Exposure of cells to 0.1 μM cadmium did not produce patterns that were significantly different from those of untreated cells. However, at 0.5 μM cadmium and above the shape parameters were different as compared to the non-treated group. It can be noted that the differences in Figure 4 appear more clearly for circularity, which is a relative value involving two parameters derived from similarly treated images (see Section 2.3), than for the skeletal length which is an absolute value that strongly depends on the way the images were filtered.

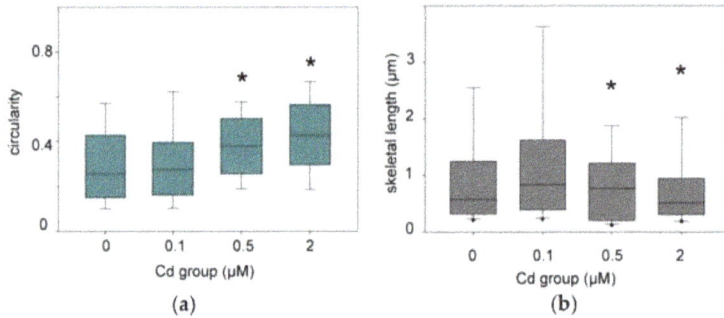

Figure 4. Box Plots showing the distribution of circularity (**a**) and skeletal lengths (**b**) of the mitochondrial network after 96 h of exposure to cadmium. * Kruskal–Wallis test $p < 0.05$ for the indicated groups compared to the reference one (0). The boundaries are the 25th and 75th percentiles and the median and error bars are plotted.

3.3. Functional Analysis of Mitochondria

3.3.1. ATP Production by INS-1 Cells after Long Term Exposure to Moderate Levels of Cadmium

One of the main functions of mitochondria is to convert energy in the form of ATP. It was thus of interest to examine whether the reorganization of the mitochondrial network observed in Section 3.2 translated into any change in the production of ATP.

The ATP/(ADP + AMP) ratio under standard growth conditions (Figure 5a) was found to decrease in a dose-dependent way, and the difference between the control group of unexposed cells and exposed ones reached statistical significance at 2.5 μM CdCl$_2$. Islets of Langerhans produce insulin in response to variations of glucose concentrations, and mitochondria are responsible for increased ATP production under these conditions. The latter was thus measured in response to increased glucose concentrations (Figure 5b). High (16.7 mM) glucose concentrations increased the ATP load of the cells as expected by a factor of approximately 50% on average as compared to low (2.7 mM) glucose concentrations. But when comparing among the cadmium treatment groups, no statistical differences were found between the energy charges of these cells.

3.3.2. Oxygen Consumption of INS-1 Cells after Long Term Exposure to Moderate Levels of Cadmium

As above for nucleotide measurements, the respiration rates were first recorded for intact cells in the complete growth medium. The basal respiration rate was compared with that of the same cells in the presence of oligomycin that inhibits the proton channel of the F_0F_1 complex V. From these data it is possible to calculate the fraction of oxygen consumption used to produce mitochondrial ATP (see Section 2.5). The plot of the results in Figure 6a shows that this fraction did not vary with the cadmium treatment below concentrations triggering cell death. Accordingly, the level of oxidative species within similarly treated cells did not change as a function of the cadmium dose as determined by reaction with dihydroethidium (see Section 2.6). The same procedure was applied to cells that were incubated in serum-free medium with either low (2.8 mM) or high (16.7 mM) glucose. As expected, respiration rates with high glucose concentrations were higher than with low concentrations, and the use of oligomycin enabled to determine which fraction of O$_2$ consumption was used for ATP synthesis. The ratio of these fractions directed to ATP under high glucose and low glucose was calculated at each cadmium concentration (Figure 6b). This ratio, which is a proxy of the coupling between glucose increase and ATP production, was found to increase in a dose-dependent way up to 1 μM, but it decreased significantly at 2.5 μM when compared to the other groups.

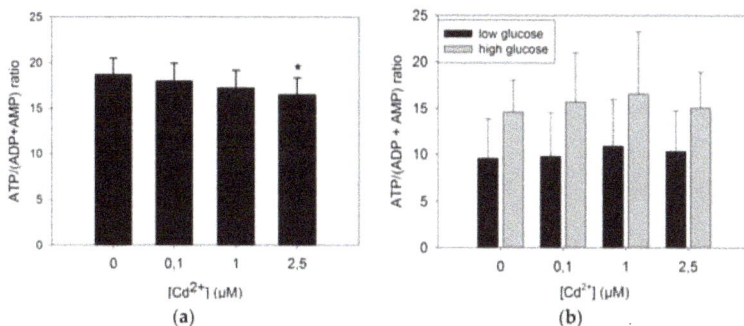

Figure 5. ATP/(ADP + AMP) ratios in INS-1 cells exposed to cadmium. (**a**) INS-1 cells were exposed to the indicated concentrations of cadmium for 72 h in the complete growth medium, and the cellular concentrations of nucleotides were measured as described in Section 2.4. A *t*-test was applied to compare each group to the non-exposed one. * $p = 0.02$, $n = 9$ for each group; (**b**) After the cadmium treatment, cells were pre-incubated for 1 h in KRBH buffer supplemented with 2.8 mM glucose, and then for another hour in the same medium (black bars) and with 16.7 mM glucose (grey bars); $n = 7$ for each group.

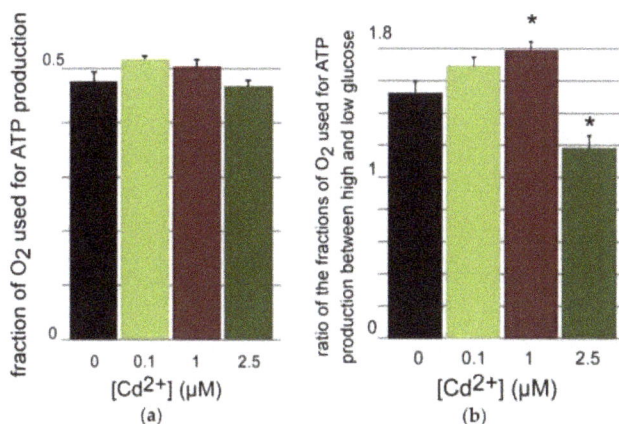

Figure 6. Respiration of INS-1 cells after 72 h-exposure to CdCl$_2$. (**a**) Respiration rates were measured as described in Section 2.5 and the slope ratios [(basal – oligomycin)/basal] were calculated for each cadmium treatment group ($n = 3$); (**b**) The same experiment was carried out in KRBH buffer supplemented with either 16.7 or 2.8 mM glucose ($n = 3$–5). The ratio of the values obtained for each series are plotted. * One-way Anova *p* values of 0.013 (1 μM) and 0.002 (2.5 μM) as compared to the control group.

4. Discussion

Considering all the data together, it appears that some characteristics of the INS-1 cell line are sensitive to sub-lethal cadmium concentrations for several generations largely before any sign of significant cellular death is detected. Under the implemented conditions, whereas the decrease of viability becomes significant only largely above 1 μM (Figure 1), the amount of cadmium accumulated over 3 days exceeds the control value by an approximate factor of 300 already at this concentration, and this factor further increases at higher concentrations (Figure 2). β-cells are very specialized in that they almost exclusively convert glucose input above the basal cellular needs into insulin production and secretion [16]. Hence mitochondria represent a first sub-cellular hub downstream of glucose absorption to generate potassium channel inhibition at the plasma membrane via increased ATP.

The prolonged exposure to sub-lethal concentrations of cadmium and the associated increase of accumulated cadmium can be correlated with a reorganization of the mitochondrial network (Figures 3 and 4). The change of mitochondrial morphology is characterized by increasingly fragmented and perinuclear mitochondria when the cadmium dose increases. Since the size of the individual mitochondria does not seem to notably change with application of cadmium, at the level of resolution afforded by the method we implemented to detect them, it may be proposed that the mitochondrial fusion-fission equilibrium is disturbed in favor of fission. Fragmented mitochondria have often been associated with impaired function [23], including in β-cells [24]. They are also observed in β-cells of Type 2 diabetic patients [25], together with altered amounts of selected mitochondrial proteins, decreased energy charge, and depolarized mitochondrial membranes. Post-fission isolated mitochondria may have a variable polarization status, but those that cannot repolarized are directed to mitophagy and loss of function [26]. Thus, the change of mitochondrial morphology that was clearly detected by the circularity parameter at the chronic concentration of 0.5 μM CdCl$_2$ and above (Figure 4) might have been expected to lead to impaired ability of the organelles to produce ATP in response to increased glucose.

But the apparent change of mitochondrial fusion-fission dynamics is differentially paralleled by functional consequences. The decreasing trend of the energetic charge, estimated by the ATP/(ADP + AMP) ratio as a function of cadmium exposure, is statistically significant only at the onset of cell death, i.e., at a 3 day-dose of 2.5 μM (Figure 5a). This result indicates that the rate of ATP production may decrease at relatively high cadmium doses, or its use is increased to counter the effects of the cadmium burden, or both. The decreasing trend of Figure 5a is observed when cells are steadily growing in the conventional growth medium containing ca. 11 mM (2 g/L) glucose. Under these conditions the O$_2$ respiration rates devoted to ATP production do not change (Figure 6a). This suggests that the decrease of the energetic charge observed in Figure 5a is due to increased ATP consumption or a small decoupling of the mitochondria in the 2.5 μM Cd-group. The relatively high respiration flow that is not used for ATP synthesis (ca. 50% in Figure 6a) is noteworthy, and it has already been noticed in β-cells models [27].

But, when glucose is decreased to 2.8 mM in minimal medium, the energy charge decreases and the difference between the 2.5 μM Cd-group and non-exposed cells cancels out. Similarly, the response to increased glucose does not change for all cadmium-treatment groups (Figure 5b). However, the stability of the energy charge among the different cadmium groups contrasts with the O$_2$ consumption devoted to ATP production which first increases as a function of cadmium exposure to reach significance at 1 μM, and then partly collapses at 2.5 μM (Figure 6b), i.e., at the onset of cell death. This phenomenon indicates that challenged cells increase the electron flow to maintain an adjusted response to increasing glucose at sub-lethal cadmium doses, but that cadmium concentrations such as 2.5 μM triggering cell death change this response and no longer request reallocation of the fraction of O$_2$ consumption used to produce ATP (Figure 6b) to respond to increasing glucose (Figure 5b).

Changes of mitochondrial morphology and function in β-cells have already been correlated with apoptosis [28,29]. But changes in morphology are not straightforwardly related to changes in glucose-stimulated insulin secretion [30]. Since β-cells can be replenished, the functional consequences of their apoptosis should occur when the β-cells mass is decreased together with insulin production [31]. The dynamics of β-cells turnover endow these cells with an efficient compensatory mechanism in line with loss of β-cells mass being a relatively late development in diabetes, of Type 2 in particular.

In a parallel way in the present study, no significant effects on the monitored parameters were observed by applying 0.1 μM cadmium for 3 or 4 days. Only a slight tendency was sometimes noticed (energetic charge, O_2 consumption upon glucose challenge) that eventually could become significant at higher cadmium concentrations. In this respect, mitochondrial dysfunction cannot be considered as an early sign of cadmium poisoning of β-cells, and any biomarker related to mitochondrial integrity is unlikely to provide a reliable and sensitive probe of mild exposure to cadmium.

These data are another illustration of one of the ways cells try to adapt to an environmental stress [32]. It is likely that the implemented molecular changes triggered by the increasing cadmium load (Figure 2) are numerous, but they do not translate into modified phenotypic traits up to relatively high concentrations ($\geq$0.5 μM). This can be compared to the situation encountered with populations exposed to environmental cadmium. They usually do not exhibit early health problems, but their chronic exposure is expected to impact molecular networks [5,33] hence contributing to morbidity and association with the development of various diseases [4].

But long-life provision of cadmium to pancreatic β-cells is likely to eventually lead to dysfunction, including of their mitochondria. The data reported herein apply to a simple cell model that has been exposed for a relatively short time as compared to the conditions that apply to long-lived animals, humans in particular. This is an obvious and important limitation of this work. However, the data suggest that non-invasive detection of cadmium in islets of Langerhans may provide a useful marker for subjects at risk of developing diabetes. Methods to detect cadmium in vivo are developing [34], and they may eventually become instrumental to probe the highly variable cadmium content of human islets [11], to be correlated with the likeliness of developing diabetes. Future work should analyze insulin secretion under the presently implemented conditions, but this single endpoint depends on additional steps lying downstream of mitochondria in the currently accepted mechanism of glucose-stimulated insulin secretion. They include plasma membrane depolarization, calcium signaling, and granules loading for instance, all events that may be sensitive to cadmium exposure, even at low concentrations. Surface receptors may also be sensitive to external cadmium and modulate the β-cells' response to increased glucose concentrations. In addition, since Cd accumulates, other intra-cellular targets may influence the overall function of these cells. Also, in line with the results reported here, the sensitivity of the mitochondrial fusion-fission proteins to the presence of cadmium should be probed. Available data indicate that a 2-way relationship between mitochondrial network dynamics and sensitivity to cadmium should be expected [35–37] along previously delineated mechanisms [5]. Furthermore, the consequences of low-level exposure of mammals to cadmium should be analyzed in details for glucose homeostasis beyond available data. This has been underway with a rat model which demonstrated that females and neonates of these laboratory animals are at risk of pre-diabetic symptoms under different modalities of chronic low level exposure to cadmium [22,38].

Acknowledgments: This work was supported by a grant from Agence Nationale de la Recherche ANR-13-CESA-008-Cadmidia. Stéphane Attia, Frédéric Lamarche, and Eric Fontaine are gratefully thanked for their advice and contributions to this work.

Author Contributions: C.D. and J.-M.M. conceived and designed the experiments; A.J., C.C.-R., J.A., K.J.S.A., R.B.M. and M.L. performed the experiments; A.J., C.C.-R., J.A., C.D. and J.-M.M. analyzed the data; J.-M.M. wrote the paper and all authors revised the manuscript.

Conflicts of Interest: The authors declare no conflict of interest. The founding sponsors had no role in the design of the study; in the collection, analyses, or interpretation of data; in the writing of the manuscript, and in the decision to publish the results.

References

1. Nawrot, T.S.; Staessen, J.A.; Roels, H.A.; Munters, E.; Cuypers, A.; Richart, T.; Ruttens, A.; Smeets, K.; Clijsters, H.; Vangronsveld, J. Cadmium exposure in the population: From health risks to strategies of prevention. *Biometals* **2010**, *23*, 769–782. [CrossRef] [PubMed]
2. Agency for Toxic Substances and Disease Registry (ATSDR). *Toxicological Profile for Cadmium*; U.S. Department of Health and Human Services: Atlanta, GA, USA, 2012; pp. 1–487.

3. European Food Safety Authority (EFSA). Cadmium dietary exposure in the european population. *EFSA J.* **2012**, *10*, 37. [CrossRef]
4. Satarug, S.; Garrett, S.H.; Sens, M.A.; Sens, D.A. Cadmium, environmental exposure, and health outcomes. *Environ. Health Perspect.* **2010**, *118*, 182–190. [CrossRef] [PubMed]
5. Moulis, J.M. Cellular mechanisms of cadmium toxicity related to the homeostasis of essential metals. *Biometals* **2010**, *23*, 877–896. [CrossRef] [PubMed]
6. Moulis, J.M.; Bourguignon, J.; Catty, P. Cadmium. In *Binding, Transport and Storage of Metal Ions in Biological Cells*; Maret, W., Wedd, A., Eds.; Royal Chemical Society: Cambridge, UK, 2014; pp. 695–746.
7. Maret, W.; Moulis, J.M. The bioinorganic chemistry of cadmium in the context of its toxicity. In *Metal Ions in Life Sciences*; Sigel, A., Sigel, H., Sigel, R.K.O., Eds.; Springer Science + Business Media B.V.: Dordrecht, The Netherlands, 2013; Volume 11, pp. 1–29.
8. Thevenod, F.; Lee, W.K. Toxicology of cadmium and its damage to mammalian organs. *Met. Ions Life Sci.* **2013**, *11*, 415–490. [PubMed]
9. Viau, M.; Collin-Faure, V.; Richaud, P.; Ravanat, J.L.; Candeias, S.M. Cadmium and T cell differentiation: Limited impact in vivo but significant toxicity in fetal thymus organ culture. *Toxicol. Appl. Pharmacol.* **2007**, *223*, 257–266. [CrossRef] [PubMed]
10. Uetani, M.; Kobayashi, E.; Suwazono, Y.; Honda, R.; Nishijo, M.; Nakagawa, H.; Kido, T.; Nogawa, K. Tissue cadmium (Cd) concentrations of people living in a Cd polluted area, Japan. *Biometals* **2006**, *19*, 521–525. [CrossRef] [PubMed]
11. El Muayed, M.; Raja, M.R.; Zhang, X.; MacRenaris, K.W.; Bhatt, S.; Chen, X.; Urbanek, M.; O'Halloran, T.V.; Lowe, W.L., Jr. Accumulation of cadmium in insulin-producing beta cells. *Islets* **2012**, *4*, 405–416. [CrossRef] [PubMed]
12. Kuo, C.C.; Moon, K.; Thayer, K.A.; Navas-Acien, A. Environmental chemicals and type 2 diabetes: An updated systematic review of the epidemiologic evidence. *Curr. Diabetes Rep.* **2013**, *13*, 831–849. [CrossRef] [PubMed]
13. Tinkov, A.A.; Filippini, T.; Ajsuvakova, O.P.; Aaseth, J.; Gluhcheva, Y.G.; Ivanova, J.M.; Bjorklund, G.; Skalnaya, M.G.; Gatiatulina, E.R.; Popova, E.V.; et al. The role of cadmium in obesity and diabetes. *Sci. Total Environ.* **2017**, *601–602*, 741–755. [CrossRef] [PubMed]
14. Wu, M.; Song, J.; Zhu, C.; Wang, Y.; Yin, X.; Huang, G.; Zhao, K.; Zhu, J.; Duan, Z.; Su, L. Association between cadmium exposure and diabetes mellitus risk: A prisma-compliant systematic review and meta-analysis. *Oncotarget* **2017**, *8*, 113129–113141. [CrossRef] [PubMed]
15. Jacquet, A.; Ounnas, F.; Lenon, M.; Arnaud, J.; Demeilliers, C.; Moulis, J.M. Chronic exposure to low-level cadmium in diabetes: Role of oxidative stress and comparison with polychlorinated biphenyls. *Curr. Drug Targets* **2016**, *17*, 1385–1413. [CrossRef] [PubMed]
16. Rorsman, P.; Braun, M. Regulation of insulin secretion in human pancreatic islets. *Annu. Rev. Physiol.* **2013**, *75*, 155–179. [CrossRef] [PubMed]
17. Martelli, A.; Rousselet, E.; Dycke, C.; Bouron, A.; Moulis, J.M. Cadmium toxicity in animal cells by interference with essential metals. *Biochimie* **2006**, *88*, 1807–1814. [CrossRef] [PubMed]
18. Chang, K.C.; Hsu, C.C.; Liu, S.H.; Su, C.C.; Yen, C.C.; Lee, M.J.; Chen, K.L.; Ho, T.J.; Hung, D.Z.; Wu, C.C.; et al. Cadmium induces apoptosis in pancreatic beta-cells through a mitochondria-dependent pathway: The role of oxidative stress-mediated c-Jun N-terminal kinase activation. *PLoS ONE* **2013**, *8*, e54374.
19. Cannino, G.; Ferruggia, E.; Luparello, C.; Rinaldi, A.M. Cadmium and mitochondria. *Mitochondrion* **2009**, *9*, 377–384. [CrossRef] [PubMed]
20. Maechler, P.; de Andrade, P.B. Mitochondrial damages and the regulation of insulin secretion. *Biochem. Soc. Trans.* **2006**, *34*, 824–827. [CrossRef] [PubMed]
21. Asfari, M.; Janjic, D.; Meda, P.; Li, G.; Halban, P.A.; Wollheim, C.B. Establishment of 2-mercaptoethanol-dependent differentiated insulin-secreting cell lines. *Endocrinology* **1992**, *130*, 167–178. [CrossRef] [PubMed]
22. Jacquet, A.; Arnaud, J.; Hininger-Favier, I.; Hazane-Puch, F.; Couturier, K.; Lénon, M.; Lamarche, F.; Ounnas, F.; Fontaine, E.; Moulis, J.-M.; et al. Impact of chronic and low cadmium exposure of rats: Sex specific disruption of glucose metabolism. *Chemosphere* **2018**. in revision.
23. Dhingra, R.; Kirshenbaum, L.A. Regulation of mitochondrial dynamics and cell fate. *Circ. J.* **2014**, *78*, 803–810. [CrossRef] [PubMed]

24. Zhang, Z.; Wakabayashi, N.; Wakabayashi, J.; Tamura, Y.; Song, W.J.; Sereda, S.; Clerc, P.; Polster, B.M.; Aja, S.M.; Pletnikov, M.V.; et al. The dynamin-related GTPase Opa1 is required for glucose-stimulated ATP production in pancreatic beta cells. *Mol. Biol. Cell* **2011**, *22*, 2235–2245. [CrossRef] [PubMed]

25. Anello, M.; Lupi, R.; Spampinato, D.; Piro, S.; Masini, M.; Boggi, U.; Del Prato, S.; Rabuazzo, A.M.; Purrello, F.; Marchetti, P. Functional and morphological alterations of mitochondria in pancreatic beta cells from type 2 diabetic patients. *Diabetologia* **2005**, *48*, 282–289. [CrossRef] [PubMed]

26. Twig, G.; Elorza, A.; Molina, A.J.; Mohamed, H.; Wikstrom, J.D.; Walzer, G.; Stiles, L.; Haigh, S.E.; Katz, S.; Las, G.; et al. Fission and selective fusion govern mitochondrial segregation and elimination by autophagy. *EMBO J.* **2008**, *27*, 433–446. [CrossRef] [PubMed]

27. Affourtit, C.; Brand, M.D. Uncoupling protein-2 contributes significantly to high mitochondrial proton leak in INS-1E insulinoma cells and attenuates glucose-stimulated insulin secretion. *Biochem. J.* **2008**, *409*, 199–204. [CrossRef] [PubMed]

28. Men, X.; Wang, H.; Li, M.; Cai, H.; Xu, S.; Zhang, W.; Xu, Y.; Ye, L.; Yang, W.; Wollheim, C.B.; et al. Dynamin-related protein 1 mediates high glucose induced pancreatic beta cell apoptosis. *Int. J. Biochem. Cell Biol.* **2009**, *41*, 879–890. [CrossRef] [PubMed]

29. Peng, L.; Men, X.; Zhang, W.; Wang, H.; Xu, S.; Xu, M.; Xu, Y.; Yang, W.; Lou, J. Dynamin-related protein 1 is implicated in endoplasmic reticulum stress-induced pancreatic beta-cell apoptosis. *Int. J. Mol. Med.* **2011**, *28*, 161–169. [PubMed]

30. Molina, A.J.; Wikstrom, J.D.; Stiles, L.; Las, G.; Mohamed, H.; Elorza, A.; Walzer, G.; Twig, G.; Katz, S.; Corkey, B.E.; et al. Mitochondrial networking protects beta-cells from nutrient-induced apoptosis. *Diabetes* **2009**, *58*, 2303–2315. [CrossRef] [PubMed]

31. Cerf, M.E. Beta cell dynamics: Beta cell replenishment, beta cell compensation and diabetes. *Endocrine* **2013**, *44*, 303–311. [CrossRef] [PubMed]

32. Moulis, J.-M. Cadmium exposure, cellular and molecular adaptations. In *Encyclopedia of Metalloproteins*; Kretsinger, R.H., Uversky, V.N., Permyakov, E.A., Eds.; Springer: New York, NY, USA, 2013; pp. 364–371.

33. Thévenod, F. Cadmium and cellular signaling cascades: To be or not to be? *Toxicol. Appl. Pharmacol.* **2009**, *238*, 221–239. [CrossRef] [PubMed]

34. Liu, Y.; Dong, X.; Sun, J.; Zhong, C.; Li, B.; You, X.; Liu, B.; Liu, Z. Two-photon fluorescent probe for cadmium imaging in cells. *Analyst* **2012**, *137*, 1837–1845. [CrossRef] [PubMed]

35. Luz, A.L.; Godebo, T.R.; Smith, L.L.; Leuthner, T.C.; Maurer, L.L.; Meyer, J.N. Deficiencies in mitochondrial dynamics sensitize *Caenorhabditis elegans* to arsenite and other mitochondrial toxicants by reducing mitochondrial adaptability. *Toxicology* **2017**, *387*, 81–94. [CrossRef] [PubMed]

36. Xu, S.; Pi, H.; Chen, Y.; Zhang, N.; Guo, P.; Lu, Y.; He, M.; Xie, J.; Zhong, M.; Zhang, Y.; et al. Cadmium induced drp1-dependent mitochondrial fragmentation by disturbing calcium homeostasis in its hepatotoxicity. *Cell Death Dis.* **2013**, *4*, e540. [CrossRef] [PubMed]

37. Xu, S.; Pi, H.; Zhang, L.; Zhang, N.; Li, Y.; Zhang, H.; Tang, J.; Li, H.; Feng, M.; Deng, P.; et al. Melatonin prevents abnormal mitochondrial dynamics resulting from the neurotoxicity of cadmium by blocking calcium-dependent translocation of drp1 to the mitochondria. *J. Pineal Res.* **2016**, *60*, 291–302. [CrossRef] [PubMed]

38. Jacquet, A.; Barbeau, D.; Arnaud, J.; Hijazi, S.; Hazane-Puch, F.; Lamarche, F.; Quiclet, C.; Couturier, K.; Fontaine, E.; Moulis, J.-M.; et al. Impact of maternal low-level cadmium exposure before and during gestation, and during lactation on metabolism of offspring at different ages. 2018; in preparation.

toxics

MDPI

Article

The NOAEL Metformin Dose Is Ineffective against Metabolic Disruption Induced by Chronic Cadmium Exposure in Wistar Rats

Victor Enrique Sarmiento-Ortega [1], Eduardo Brambila [1], José Ángel Flores-Hernández [1], Alfonso Díaz [2], Ulises Peña-Rosas [3], Diana Moroni-González [1], Violeta Aburto-Luna [1] and Samuel Treviño [1,*]

[1]   Laboratory of Chemical-Clinical Investigations, Department of Clinical Chemistry, Faculty of Chemistry Science, University Autonomous of Puebla, 14 South. CQ1, University City, Puebla C.P. 72560, Mexico; qfb_veso111@hotmail.com (V.E.S.-O.); eduardobrambila1@yahoo.com.mx (E.B.); quimicoangel32@hotmail.com (J.Á.F.-H.); d.moroni_25@hotmail.com (D.M.-G.); val_140485@hotmail.com (V.A.-L.)
[2]   Department of Pharmacy, Faculty of Chemistry Science, University Autonomous of Puebla, 14 South. CQ1, University City, Puebla C.P. 72560, Mexico; dan_alf2005@yahoo.com.mx
[3]   Department of Analytic Chemistry, Faculty of Chemistry Science, University Autonomous of Puebla, 14 South. CQ1, University City, Puebla C.P. 72560, Mexico; quim_perua@hotmail.com
*   Correspondence: samuel_trevino@hotmail.com; Tel.: +521-222-229-5500 (ext. 7367)

Received: 16 August 2018; Accepted: 7 September 2018; Published: 10 September 2018

Abstract: Previous studies have proposed that cadmium (Cd) is a metabolic disruptor, which is associated with insulin resistance, metabolic syndrome, and diabetes. This metal is not considered by international agencies for the study of metabolic diseases. In this study, we investigate the effect of metformin on Cd-exposed Wistar rats at a lowest-observed-adverse-effect level (LOAEL) dose (32.5 ppm) in drinking water. Metabolic complications in the rats exposed to Cd were dysglycemia, insulin resistance, dyslipidemia, dyslipoproteinemia, and imbalance in triglyceride and glycogen storage in the liver, muscle, heart, kidney, and adipose tissue. Meanwhile, rats treated orally with a No-observable-adverse-effect level (NOAEL) dose of metformin (200 mg/kg/day) showed mild improvement on serum lipids, but not on glucose tolerance; in tissues, glycogen storage was improved, but lipid storage was ineffective. In conclusion, metformin as a first-line pharmacological therapy must take into consideration the origin and duration of metabolic disruption, because in this work the NOAEL dose of metformin (200 mg/kg/day) showed a limited efficiency in the metabolic disruption caused by chronic Cd exposure.

Keywords: metformin; cadmium toxicity; metabolic disruptor; metabolic syndrome

1. Introduction

Cadmium (Cd) is a transition metal that represents a health risk, being classified as one of the top five most hazardous environmental contaminants by the Agency for Toxic Substances and Disease Registry [1]. Human exposure to Cd occurs mainly through inhalation or ingestion, and its absorption depends on the particle size, concentration, time-exposure, and competitivity with biometals such as iron, calcium, or zinc. Cigarette smoking is considered to be the most significant source of human exposure to Cd [2–5]. In humans and other mammals, Cd can damage several organs and tissues, including the kidneys, liver, lung, pancreas, testis, placenta, brain, and bone, but the kidneys and liver are the two primary target organs [5–8]. Damage to tissues is accompanied by a variable degree of injury because of inflammation and oxidative stress [9–13]. Likewise, Cd is referred to as a heavy metal that causes endocrine disruption [14–16], and recently as a metabolic disruptor because it has been described as a risk factor for the developing of insulin resistance, metabolic syndrome, obesity, and diabetes; however, the international agencies for the study of metabolic diseases have not yet considered these issues officially [17–20].

According to world sanitary statistics in 2014 emitted by the World Health Organization (WHO), there are almost 387 million diagnosed cases of type 2 diabetes mellitus (T2DM), but it is estimated that 178 million remain undiagnosed and this is expected to reach 592 million in 2035, which will contribute to health expenses of approximately $245 billion in the U.S. alone. Correspondingly, diabetes has been intimately linked to obesity and overweight problems, which represent a third of the worldwide population. Because of its extremely high prevalence, obesity has a significant socioeconomic impact of approximately $190 billion/year in the U.S. [21–24]. Obesity and T2DM belong to a very complex group of genetic and epigenetic diseases with a socio-environmental influence known as chronic non-communicable diseases, that have a common background: metabolic disturbances or metabolic syndrome associated with dysglycemia, dyslipidemia, dyslipoproteinemia, and arterial hypertension, as well as hormone imbalance of insulin, leptin, adiponectin, and resistin, which affect other hormonal axes, contributing to alterations of triglycerides and glycogen in several tissues [25–27].

The first line of pharmacological therapy for metabolic disorders is metformin (1,1-dimethyl biguanide) because it can control each complication associated with metabolic syndrome in variable degrees. With approximately 50 years of accumulated global clinical experience, metformin is generally regarded as safe [28]. Metformin has demonstrated its efficiency in lowering blood glucose levels, reducing mild weight problems in people with a high body mass index (BMI), improving insulin sensitivity and insulin secretion, and modulating multiple incretin axis components, all of which have only a minimal risk of hypoglycemia, and regulating triglyceride and cholesterol levels [29,30]. Recently, metformin has been confirmed by the American Diabetes Association and the European Association for the Study of Diabetes as a pharmacological therapy [31,32]. However, the dosage is a sensitive issue because the diabetic patients can consume up to 2000 mg per day, in two to three divided doses. In this sense, the adaptation of therapeutical doses has been studied in animal models, such as rats, in order to understand the toxicological effects. The hypoglycemic effects with a no observable adverse effect level (NOAEL) were 200 mg/kg/day. Meanwhile, a dose of $\geq$600 mg/kg/day observed adverse findings including an increased incidence of minimal necrosis, inflammation, and metabolic acidosis (increased serum lactate and beta-hydroxybutyric acid and decreased serum bicarbonate and urine pH); a dose of $\geq$900 mg/kg/day resulted in moribundity/mortality and clinical signs of toxicity [33]. Although the therapeutic dose has been deeply studied, the exact mechanism of action for metformin is still not completely understood, but it is known that in various tissues and organs, it improves glucose metabolism via activation of the ubiquitously expressed AMP-activated protein kinase (AMPK) [17,34]. The AMPK is a Ser/Thr protein kinase that acts as a sensor of the cellular energy status and modulates metabolic pathways of carbohydrates and lipids via the inhibition of enzymes involved in gluconeogenesis and glycogen synthesis. Thus, overproduction of glucose from the liver is controlled by means of decreasing the phosphorylation of essential substrates for glucose output, reducing cAMP and glucagon action, as well as AMPK activation in a fasting state and inhibiting fatty acid synthesis, while mitochondrial oxidative phosphorylation is stimulated [35,36].

Therefore, due to a LOAEL dose of Cd causes metabolic disruptions such as insulin resistance, dyslipidemia, dysglycemia, and metabolic syndrome. The aim of this work was to investigate, in Wistar rats, the effect of a NOAEL dose of metformin on the homeostasis of carbohydrates and lipid in serum and tissues after a chronic Cd exposition.

2. Material and Methods

2.1. Animals and Treatments

One hundred male Wistar rats, weighing 70 to 80 g, obtained from the Claude Bernard vivarium of the Universidad Autónoma de Puebla, Mexico were housed in polycarbonate boxes with a sawdust bed and maintained under temperature-controlled (19–26 °C), 12-h light–dark cycles, with free access to food and water. The animals were conditioned with a standard diet until reaching 100 g. Once the rats reached this weight, they were randomly divided into two groups: "Control" (standard diet Labdiet 5001 and water *ad libitum*, n = 50) and "Cadmium" (standard 5001 Labdiet diet with drinking water containing 32.5 ppm of Cd *ad libitum*, n = 50). At the end of the third month, 10 "Control" rats and 10 rats from the "Cadmium" group were sacrificed to ensure the metabolic disruption. The cadmium group was then divided into two subgroups: "Cadmium" alone subgroup (n = 20), and the "Cd + Metformin" subgroup (standard 5001 Labdiet diet, drinking water containing 32.5 ppm of Cd *ad libitum* and Metformin treatment 200 mg/kg/day; oral via; n = 20). The control group also was divided into control (standard 5001 Labdiet diet, drinking water free Cd; n = 20) and Metformin (standard 5001 Labdiet diet, drinking water free Cd and Metformin treatment 200 mg/kg/day; oral via; n = 20) groups. All groups were kept under these conditions for two more months. The metformin dose used was chosen based on previous reports of a no observable adverse effect level (NOAEL) and the effective dose (ED$_{50}$) as hypoglycemic and hypolipidemic [25,33]. Just prior to each cohort time (3, 4, and 5 months), the rats received an oral glucose load (TOG), equivalent to 1.75 g of glucose/kg weight. The rats were anesthetized intraperitoneally with xylazine/ketamine (20/137 mg/kg) and under anesthesia, whole blood (500 μL) was drawn via cardiac puncture at 0, 30, 60, and 90 min. The serum was then separated by centrifugation and stored at −70 °C, and after, tissues (liver, biceps femoris muscle, heart, kidney, and retroventral adipose tissue) were immediately removed and thoroughly perfused with cold saline and stored at −70 °C until the analysis. Each procedure was performed according to the National Institute of Health's guide for the care and use of Laboratory Animals and the Guide for the Care and Use of Laboratory Animals of the Mexican Council for Animal Care NOM-062-ZOO-1999, European Convention for the Protection of Vertebrate Animals Used for Experimental and other Scientific Purposes, Guiding Principles in the Use of Animals in Toxicology, and it was approved by the Institutional Committee for the Care and Use of Animals on 10 November 2015. Every effort was made to minimize the number of animals used and to ensure minimal pain and/or discomfort.

2.2. Animal Zoometry

Weight, fat percentage, and size of the rats were monitored weekly. The weight was measured using a digital balance (Torrey, City of Mexico, State of MEX, Mexico; model: LPCR-20/40) and the size of each animal was obtained by measuring the length from the base of the tail to the tip of the nose. The abdomen diameter was estimated using the diaphragm zone as an upper limit and the fold of the legs as the bottom limit. The body mass index (BMI) was calculated using the formula weight/size^2 and fat percentage was calculated according to the Lee index for rodent models, with the formula: % fat = [(weight in g $^{(0.33)}$)/size in cm] × 100 [37].

2.3. Biochemical Assays in Serum

From the serum obtained at time 0 min after 4–5 h fasting, the concentrations of glucose, lactate, total lipids, triglycerides, cholesterol, low-density lipoprotein cholesterol (LDL), and high-density lipoprotein cholesterol (HDL) were determined using spectrophotometry with commercial kits and an automatic analyzer AutoKemII (KONTROLab, Company, Morelia, MICH, Mexico). The level of very low-density lipoprotein (VLDL) was obtained using the Friedenwald equation [38]. Free fatty acid (FFA) concentration was determined according to the method described by Brunk and Swanson (1981), in a Perkin Elmer EZ150 model Lambda (Tres Cantos, MAD, Spain) spectrophotometer at 620 nm wavelength [39]. Lipoprotein sub-fractions were characterized using a polyacrylamide gel disc electrophoresis, as described by Rainwater et al. [40]. Three gradients of different pore size were prepared to allow for the separation of pre-beta (VLDL1 and VLDL2), beta (LDL I, II, IIIA, IIIB, IVA, and IVB) and alpha (HDL2a, 2b, 3a, 3b, and 3c) sub-fractions. To determine the different levels of lipoprotein sub-fractions, a densitometric analysis of the discs was performed in the polyacrylamide gel and then the area under the curve was quantified by using ImageJ software (National Institutes of Health, Bethesda, MD, USA).

2.4. Insulin Resistance Analysis

Plasma insulin concentrations were determined using an ELISA immunoassay (Diagnostica Internacional Company, Guadalajara, JAL, Mexico), with the resulting antibody–antigen complex assessed at 415 nm in a Stat fax 2600 plate reader (WinerLab Group, ROS, Argentina). Insulin resistance using homeostasis model assessment insulin resistance (HOMA-IR), insulin resistance adipocyte dysfunction (IDA-IR), and insulin sensitivity using hepatic insulin sensitivity (HIS) was evaluated using mathematical models according to the report by Treviño et al. [37].

2.5. Glycogen and Triglycerides Content in Tissues

Biopsies from tissues (liver, heart, renal cortex, renal medulla, and retroventral adipose) were homogenized at 100 mg in 800 µL of isotonic saline solution (ISS) to assess triglyceride content, whereas a second dilution was made only for adipose tissue, in which the homogenate was diluted 1:2 with ISS and the protocol for the triglyceride kit described by the manufacturer was followed. For the determination of glycogen, we followed the technique described by Bennett et al., from 150 mg of each tissue homogenized with 2 mL of perchloric acid [41].

2.6. Statistical Analysis

The results are expressed as a mean ± standard error of the mean (SEM) before beginning the metformin treatment (3 months). The statistical difference between the control and the cadmium group was determined by using a Student unpaired t-test with a significance level of $p \leq 0.05$. Results obtained after 4 and 5 months of treatments were analyzed by using a one-way ANOVA test and Bonferroni post hoc test, considering $p \leq 0.05$ as statistically significant.

3. Results

3.1. Morphometry, Lipids and Carbohydrates in Serum and Tissues after 3 Months of Cd Exposure

The chronic Cd exposure in a lowest observed adverse effect level dosage (LOAEL, 32.5 ppm) after 3 months produced zoometry modifications that increased weight (29%), abdominal perimeter (22%), body mass index (45%), and percentage of fat (15%). The lipid profile showed a similar result by significantly increasing total lipids (32%) and triglycerides (138%) in serum. Although the total cholesterol showed no difference, the VLDL and LDL fractions increased by 42% and 98%, respectively. Meanwhile, the HDL fraction showed a significant decrease of 45% (Table 1). The analysis of subfractions (Figure 1) showed an increase in V1 (35%) and V2 (26%), as well as LDL I, II, IIIa, IIIb, IVa, and IVb subfractions of 12%, 12%, 13%, 20%, 26%, and 23%, respectively. Exclusively, the HDL3c subfraction increased by 31% in comparison to the control group (Figure 1). Dyslipidemia from Cd exposure also affects triglycerides stored in different tissues, increasing significantly in the liver (39%), muscle (198%), heart (78%), renal cortex (112%), renal medulla (54%) and retroventral adipose (27%).

On the other hand, the carbohydrate homeostasis was also affected, fasting glucose and postprandial glucose after a load of 1.75 g/kg showed significant increases, which corresponded to 106%, 125% (30′ post-load), 193% (60′ post-load), and 210% (90′ post-load); likewise, the lactate level was augmented by 28%. However, in some tissues, glycogen deposits were significantly diminished: in the liver (36%), heart (37%), and renal cortex (53%), while the muscle showed an increase of 107%. The glycogen content in the renal medulla and retroventral adipose showed no difference (Table 1). The zoometric, metabolic, and biochemical changes observed in the rats of the Cd group were in concordance with significant hyperinsulinemia (75%), the development of insulin resistance demonstrated by HOMA-IR (216%), and insulin resistance adipocyte dysfunction (IDA-IR; 557%), as well as a significant loss of hepatic insulin sensitivity (HIS; 72%) (Table 1).

Table 1. Metabolic disturbances caused by cadmium exposure.

Measurements	Control n = 30	Cadmium n = 50	Metabolite mg/100 mg of Tissue	Control n = 10	Cadmium n = 10
Morphometric panel:			**Triglycerides:**		
Weight (g)	341 ± 3.4	439.2 ± 8.1 *	Liver	16.2 ± 0.7	22.5 ± 1 *
Abdominal perimeter (cm)	18.4 ± 0.1	22.4 ± 0.2 *	Muscle	3.36 ± 0.2	10.0 ± 0.3 *
Body mass index	0.93 ± 0.02	1.35 ± 0.05 *	Heart	5.06 ± 0.3	9 ± 0.2 *
% Body fat	35.9 ± 0.1	41.3 ± 0.5 *	Renal cortex	5.98 ± 0.4	12.7 ± 0.6 *
Lipidic panel (mg/dL):			Renal medulla	7.46 ± 0.3	11.5 ± 0.7 *
Total lipids	184 ± 12	243 ± 5.1 *	Rv Adipose	51.2 ± 1.4	64.8 ± 1.6 *
FFA	2.18 ± 0.03	5.18 ± 1.1 *			
Triglycerides	64.4 ± 2.5	106.8 ± 3.1 *	**Glycogen:**		
Total Cholesterol	103.1 ± 7.5	103.4 ± 6	Liver	4.2 ± 0.4	2.9 ± 0.3 *
Cholesterol fraction:			Muscle	0.3 ± 0.03	0.62 ± 0.1 *
VLDL	13.5 ± 1.5	19.2 ± 1.2 *	Heart	0.9 ± 0.05	0.57 ± 0.8 *
LDL	24.2 ± 4	47.9 ± 3.1 *	Renal cortex	1.2 ± 0.2	0.57 ± 0.11 *
HDL	65.4 ± 2	36.3 ± 1.5 *	Renal medulla	0.56 ± 0.2	0.51 ± 0.6
Carbohydrate panel:			Rv. Adipose	0.55 ± 0.03	0.48 ± 0.01
Lactate (mmol/L)	7.3 ± 0.7	9.4 ± 0.5 *	**Insulin resistance panel:**		
Fasting glucose (mg/dL)	80 ± 3.2	165 ± 5 *	Insulin (µUI/mL)	12 ± 3.1	21 ± 4.5 *
‡ Glucose 30 min (mg/dL)	107.1 ± 2.9	241 ± 8.1 *	HOMA-IR	0.44 ± 0.05	1.39 ± 0.19 *
‡ Glucose 60 min (mg/dL)	90.6 ± 3.4	265 ± 7 *	IDA-IR	0.07 ± 0.03	0.46 ± 0.15 *
‡ Glucose 90 min (mg/dL)	81.2 ± 3	251.5 ± 4.2 *	HIS	18.8 ± 3.5	5.2 ± 2.2 *

The biochemical and morphometric results shown are the average of 80 separate experimental animals ± SEM. Meanwhile, the metabolite/100 mg of tissue results shown are 20 separate experimental animals ± SEM. (*) Indicates significant differences from the control group with $p \leq 0.05$ using a Student *t*-test. (‡) values obtained after a load glucose 1.75 g/kg. FFA = free fatty acid; VLDL = very low-density lipoprotein; LDL = low-density lipoprotein; HDL = high-density lipoprotein; Rv Adipose = retroventral adipose. HOMA-IR = homeostasis model assessment insulin resistance; IDA-IR = Insulin resistance adipocyte dysfunction; HIS = Hepatic insulin sensitivity.

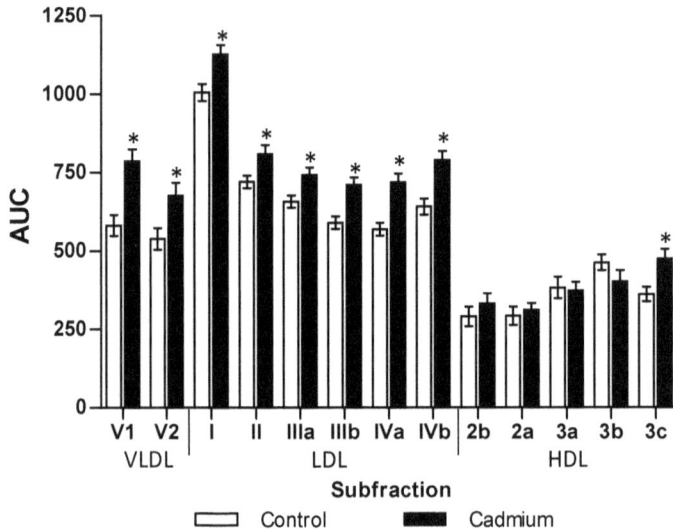

Figure 1. Disturbances caused by cadmium exposition on subfractions of lipoproteins. The results shown are the average of ten separate experimental animals ± SEM. (*) Indicates significant differences from the control group with $p \leq 0.05$ using a Student *t*-test.

3.2. Metformin Treatment on the Metabolic Disruption Caused by Cd Exposure

The metformin group after 1 and 2 months of administration did not show differences in the zoometry and serum parameters in relation to the control group. The Cd group remained altered after the fourth month of exposition in zoometric and biochemical parameters (Table 2). Meanwhile, the Cd + metformin group (1 month of treatment) showed an improvement of zoometric parameters, such as weight, abdominal perimeter, and BMI; however, there was now a complete regulation in the percentage of fat, which remained 6% above the control group (Table 2). The lipid biomarkers also showed a slight improvement without reaching the values of the control group, where total lipids, FFA, and triglycerides remained high at 24%, 79%, and 66%, respectively. However, the lipoproteins VLDL, LDL, and HDL improved. According to the total fractions, the subfractions of VLDL (V1 and V2) showed no differences compared with control group, while the subfractions of LDL I to IVa (25%, 20%, 12%, 8%, and 18%) and the HDL 3a–3c (17%, 14%, and 30%) (Figure 2) were significantly diminished. All subfractions of the Cd + metformin group showed improvement in relation to the Cd group. Triglycerides in the tissues of the Cd + metformin group showed a significant reduction compared to the Cd group, except for the adipose tissue. However, when the Cd + metformin group was compared with the control group, triglycerides remained high in the liver (18%), muscle (201%), heart (87%), renal cortex (89%), renal medulla (100%), and retroventral adipose tissue (55%) (Figure 3B). The metformin group did not show differences of stored triglycerides in tissues versus the control group.

Table 2. Zoometric and metabolic evaluation after 1 month of metformin treatment.

Measurements	Control $n = 10$	Metformin $n = 10$	Cadmium $n = 10$	Cd + Metformin $n = 10$
Morphometric panel:				
Weight (g)	401.6 ± 7.9	380.2 ±15.3	470 ± 6.3 *	418 ± 10 ▼
Abdominal perimeter (cm)	20.8 ± 0.1	19.4 ± 0.6	24.2 ± 0.7 *	21.1 ± 0.4 ▼
Body mass index	1.1 ± 0.04	1.0 ± 0.02	1.4 ± 0.01 *	1.1 ± 0.03 ▼
% Body fat	37.8 ± 0.3	35.7 ± 1.6	42.1 ± 0.1 *	40.0 ± 0.3 *▼
Lipid panel (mg/dL):				
Total lipids	187.2 ± 9.6	180.1 ± 3.3	253.6 ± 4.5 *	232 ± 6.1 *▼
FFA	2.85 ± 0.2	3.01 ± 0.2	6.49 ± 0.1 *	5.1 ± 0.1 *▼
Triglycerides	56 ± 6	51 ± 2	112.3 ± 4.5 *	93.1 ± 3.5 *▼
Total Cholesterol	111.9 ± 5.6	107.9 ± 4.9	108.1 ± 6.7	84.9 ± 5.7 * ▼
Cholesterol fraction				
VLDL	16 ± 1.6	18 ± 0.9	21.2 ± 0.9 *	18.2 ± 1 *▼
LDL	36 ± 3.5	40 ± 6.1	54.8 ± 1.7 *	26 ± 2.1 *▼
HDL	59.9 ± 2.6	49.9 ± 7.2	32.1 ± 2.5 *	40.7 ± 3.1 *▲
Carbohydrate panel:				
Lactate (mmol/L)	7.3 ± 0.7	7.5 ± 0.2	8.55 ± 0.3 *	8.2 ± 0.1 ▼
Fasting glucose (mg/dL)	80.0 ± 6.3	76.0 ± 4.8	135 ± 4.1 *	79.7 ± 4.3 ▼
Glucose 30 min (mg/dL)	107.1 ± 3.8	97.7 ± 5.2	238 ± 6.7 *	154.4 ± 5.4 *▼
Glucose 60 min (mg/dL)	90.6 ± 4.3	88.4 ± 7.1	250 ± 9.4 *	160.4 ± 3.8 *▼
Glucose 90 min (mg/dL)	81.2 ± 5.0	73.9 ± 9.4	235 ± 8.8 *	142.2 ± 6.1 *▼
Insulin resistance panel:				
Insulin (μUI/mL)	10 ± 3.4	9.8 ± 4.8	19 ± 2.4 *	27 ± 4.9 *
HOMA-IR	0.44 ± 0.03	0.40 ± 0.08	1.42 ± 0.19 *	0.88 ± 0.21 *▼
IDA-IR	0.02 ± 0.02	0.02 ± 0.01	0.54 ± 0.12 *	0.35 ± 0.11 *
HIS	22.5 ± 5.19	20.1 ± 3.3	7.01 ± 3.8 *	8.3 ± 2.2 *

The results shown are the average of 10 separate experimental animals ± SEM. (*) Indicates significant differences from the control group $p \leq 0.05$. (▼) Indicate significant decreases with respect to the cadmium group. (▲) indicates significant increases with respect to the cadmium group $p \leq 0.05$ using a one-way ANOVA test with a Bonferroni post hoc test.

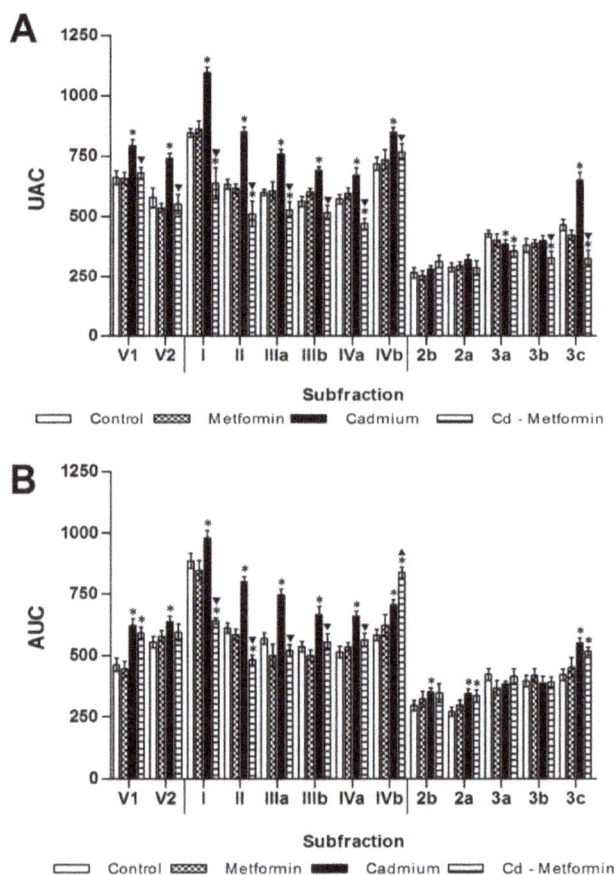

Figure 2. Metformin effect on lipoprotein subfractions. (**A**) One month with the different treatments. (**B**) Two months with the different treatments. The results shown are the average of ten separate experimental animals ± SEM. (*) Indicates significant differences from the control group. (▼) Indicates significant decreases with respect to the cadmium group. (▲) Indicates significant increases with respect to the cadmium group $p \leq 0.05$ using an ANOVA test with a Bonferroni post hoc test.

In relation to the glucose homeostasis, Cd exposure affected the oral glucose tolerance and glycogen concentration while decreasing it in the tissues (except in muscle; Figure 3A). Metformin administration did not affect lactate level, the oral glucose tolerance and increased the glycogen level in the liver, muscle, and heart. Also, metformin co-administered with Cd mild improved the oral glucose tolerance. In this regard, lactate and fasting glucose did not show differences in relation to the control group (Table 2), although glucose remained elevated postprandially, with 44% (30′ post-load), 77% (60′ post-load), and 75% (90′ post-load). Moreover, the glycogen level in the liver and retroventral adipose tissue was no different in comparison with the control group but was higher in muscles (134%), heart (60%), renal cortex (32%), and renal medulla (92%) (Figure 3A). Compared to the control group, both the Cd group (90%) and the Cd + metformin group (170%) presented hyperinsulinemia, although, in the Cd + metformin group, HOMA-IR and ADA-IR improved, although not as much as in the control group, and HIS remained the same (Table 2).

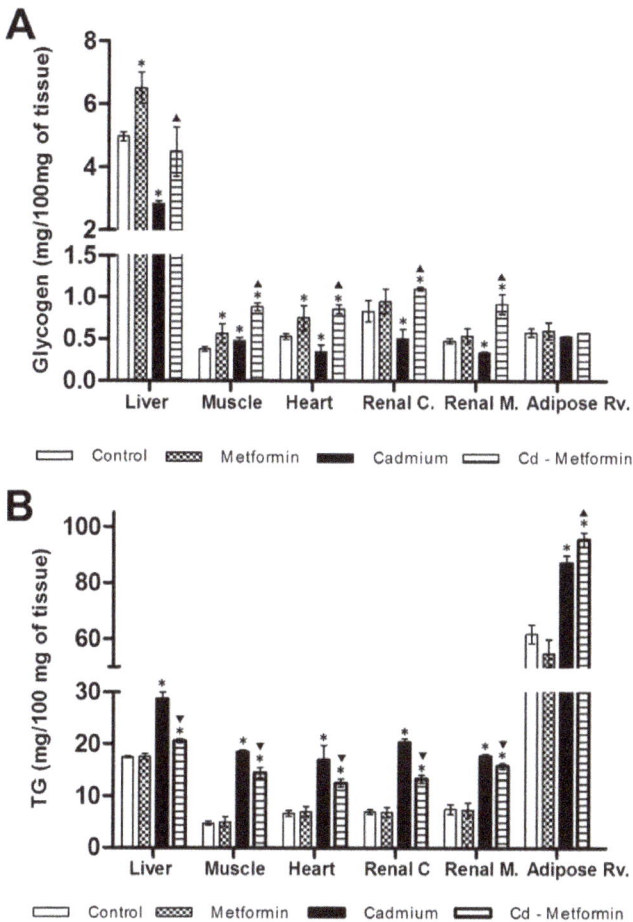

Figure 3. Glycogen and triglyceride concentration in tissues at one month with the different treatments: (**A**) glycogen, and (**B**) triglycerides. The results shown are the average of ten separate experimental animals ± SEM. (*) Indicates significant differences from the control group. (▼) Indicates significant decreases with respect to the cadmium group. (▲) Indicates significant increases with respect to the cadmium group $p \leq 0.05$ using an ANOVA test with a Bonferroni post hoc test.

After five months, rats exposed to Cd presented a greater metabolic disorder, showing evidence of zoometric worsening as well as lipid and glucose homeostasis in both serum and tissues, with a marked development of insulin resistance and loss of hepatic insulin sensitivity. In contrast, animals exposed to Cd and treated with metformin for two months showed zoometric parameters like the control group, except for the percentage of the body fat (5% greater). The panel of lipids showed incomplete recovery because the levels of total lipids remained increased (44%), FFA (69%), triglyceride (125%), and VLDL (141%), while HDL remained low (35%) (Table 3). Cholesterol subfractions of the metformin-treated group exhibited levels like the control group in V2 but not in V1 (28%, high), in LDL I to IVa but not in IVb (44%, high), as well as HDL2b, 3a, and 3b, but not in 2a and 3c (24% and 23%, high) (Figure 2B). The triglyceride deposits in rats administrated with metformin alone showed a tendency to increase; in this group, the heart observed a significant difference. With regard to the group co-treated with Cd and metformin, it showed overstoring in the liver (91%), in the muscle (384%), heart (176%), and renal cortex and medulla (451% and 369%), and retroventral adipose (21%) (Figure 4B).

Table 3. Zoometric and metabolic evaluation after 2 months of metformin treatment.

Measurements	Control $n = 10$	Metformin $n = 10$	Cadmium $n = 10$	Cd + Metformin $n = 10$
Morphometric panel:				
Weight (g)	434.3 ± 6.2	420.9 ± 12.8	502.6 ± 9.5 *	442.6 ± 10.5 ▼
Abdominal perimeter (cm)	22.1 ± 0.6	20.7 ± 0.9	26.2 ± 0.5 *	22.9 ± 0.6 ▼
Body mass index	1.1 ± 0.04	1.0 ± 0.08	1.5 ± 0.03 *	1.2 ± 0.03 ▼
% Body fat	38.8 ± 0.3	36.2 ± 1.1	43.1 ± 0.1 *	40.8 ± 0.3 *▼
Lipid panel (mg/dL):				
Total lipids	180 ± 13	178 ± 9	270 ± 6.1 *	259 ± 4.5 *
FFA	3.49 ± 0.12	3.98 ± 0.6	6.8 ± 0.16 *	5.9 ± 0.2 *▼
Triglycerides	50.5 ± 5.8	47.7 ± 5.1	121.7 ± 2.1 *	113.6 ± 3.1 *▼
Total Cholesterol	102.2 ± 10.8	98.5 ± 6.7	112.9 ± 5.2	102.6 ± 3.2
Cholesterol fraction				
VLDL	10.1 ± 1.8	10.1 ± 1.8	24.9 ± 1.2 *	24.3 ± 1.2 *
LDL	39 ± 7	41 ± 3	59.5 ± 2.7 *	43.5 ± 4.7 ▼
HDL	53.1 ± 2	50.9 ± 3	28.5 ± 2.4 *	34.8 ± 1.4 *▲
Carbohydrate panel:				
Lactate (mmol/L)	8.2 ± 0.5	9.1 ± 1.3	11.3 ± 0.4 *	10.1 ± 0.3 *▼
Fasting glucose (mg/dL)	72.9 ± 4	67.7 ± 7	139.5 ± 4.2 *	85 ± 5.9 *▼
Glucose 30 min (mg/dL)	93.1 ± 0.6	85.3 ± 4.9	157.4 ± 14.4 *	108.3 ± 6.0 *▼
Glucose 60 min (mg/dL)	98.2 ± 3.0	104.4 ± 5.5	171.3 ± 8.7 *	117.0 ± 5.3 *▼
Glucose 90 min (mg/dL)	90.0 ± 5.0	98.0 ± 7.3	164 ± 6.3 *	116.2 ± 4.8 *▼
Insulin resistance panel:				
Insulin (µUI/mL)	11 ± 1.8	14 ± 2.9	25 ± 3.2 *	34.7 ± 4.1 *▲
HOMA-IR	0.47 ± 0.03	0.50 ± 0.1	1.69 ± 0.2 *	1.21 ± 0.23 *
IDA-IR	0.02 + 0.01	0.03 ± 0.01	0.63 ± 0.17 *	0.51 ± 0.2 *
HIS	22.4 ± 4.7	24.7 ± 5.1	6.02 ± 1.9 *	6.10 ± 2.1 *

The results shown are the average of 10 separate experimental animals ± SEM. (*) Indicates significant differences from the control group $p \leq 0.05$. (▼) Indicates significant decreases with respect to the cadmium group. (▲) Indicates significant increases with respect to the cadmium group $p \leq 0.05$ using a one-way ANOVA test with a Bonferroni post hoc test.

Regarding glycogen, the metformin treatment itself increased levels in the liver, muscle, renal cortex, and heart, but the heart increase was not significant. Meanwhile, in rats Cd-exposed co-administered with metformin glycogen improved the content in the liver (6%), muscle (157%), heart (48%), and renal medulla (12%), but not in the renal cortex, where it diminished (36%) (Figure 4A). Likewise, lactate and the oral glucose tolerance showed improvement after metformin treatment, although it remained slightly elevated (0′, 17%; 30′, 16%; 90′, 19%; 90′, 29%). Hyperinsulinemia was also observed (216%), in concordance with insulin resistance by HOMA (157%) and IDA-IR (2450%), both greater than the control group, while hepatic insulin sensitivity fell (73%) (Table 3).

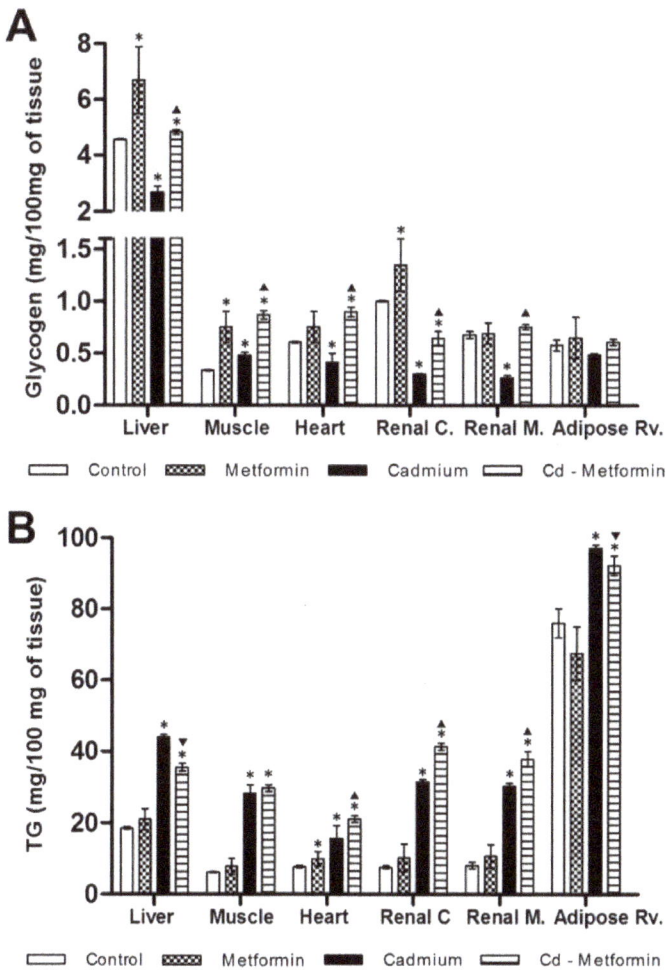

Figure 4. Metformin effect on lipoprotein subfractions. (**A**) Two months with the different treatments. (**B**) Two months with the different treatments. The results shown are the average of 10 separate experimental animals $\pm$ SEM. (*) Indicates significant differences with values above the control group. (▼) Indicates significant decreases with respect to the cadmium group. (▲) Indicates significant increases with respect to the cadmium group $p \leq 0.05$ using an ANOVA test with a Bonferroni post hoc test.

4. Discussion

In this paper, we studied the role of metformin after a metabolic disruption caused by exposure to Cd. Previously, we demonstrated that a LOAEL dose of Cd in drinking water produces a metabolic toxicity that is characterized by insulin resistance in multiple peripheral tissues, increasing insulin release with hyperglycemia and lipid metabolism alterations [37]. Although Cd toxicology has been extensively discussed, associating its toxic effects with inflammation, oxidative stress, and genotoxicity processes [2,42], the metabolic toxicity is not considered, and thus, the associated mechanism is poorly studied. The literature is somewhat contradictory in relation to Cd exposure and metabolic complications in lipids, glucose, overweight, obesity, and diabetes [43–46]. The results obtained in this work show clearly that a chronic exposition to a Cd LOAEL dose produces the metabolic disorders mentioned above. In addition to the metabolic changes, our results demonstrate new evidence of the progressive accumulation of triglycerides in different tissues and disturbances on glycogen deposits.

In humans, exposure to Cd has been related to hyperglycemia, impaired fasting glucose, and a positive correlation in patients with type 2 diabetes mellitus in a dose-dependent manner related to the kind and time of exposure [47–49]. In animal models, the hyperglycemia produced by Cd has been related with significant increases in the hepatic transporter GLUT2, carbohydrate regulatory element binding protein (ChREBP), and mRNAs of glucokinase and pyruvate kinase [50], as well as by a downregulation of the expression of the glucose transporter GLUT4 in both muscle and adipocytes. These changes lead to a limited glycolysis, increases in the glycogenolysis, and the enzymatic activation of the gluconeogenesis pathway, which could explain the lactate increase and the poor glycogen storage observed in this and other works with rats exposed to cadmium [51–55]. It is well known that hyperglycemia is normally compensated with hyperinsulinemia as an adaptive response from the pancreas to restore glucose homeostasis, which is often linked to progressive insulin resistance, a key factor in the feedback between the liver and adipose tissue in relation to triglyceride storage. However, a limited number of works have shown the insulin resistance in relation to Cd exposure [37,56,57].

Hepatic insulin resistance, or low insulin sensitivity, favors lipogenesis, and thus the increase of novo synthesis of triglycerides and a higher secretion of VLDL type V1 and V2 (triglyceride-rich lipoproteins, TRLs), as was observed in our results [57–59]. VLDL carry triglycerides, increasing its accumulation in peripheral tissues (Table 1). Mechanisms associated with insulin resistance in adipose tissue, provoke an erroneous triglyceride storage, and thus an over-flux of FFA toward all peripheral tissues, the liver being the most affected such that more VLDL of type V1 is synthesized. These events produce a redundant cycle in which high levels of TLR and FFA in serum can be observed. TRLs induce small LDL formation that could contribute to atherosclerosis development and cardiovascular disease. Small LDL subfractions, corresponding to LDL-IIIa, IIIb, IVa, and IVb, are increased in the Cd-exposed rats [60–62]. TLRs could also alter HDL subfractions, where the HDL3c subfraction is particularly susceptible to Cd exposure. Under normal conditions, HDL3c acts as an anti-oxidant against LDL oxidation and has anti-thrombotic, anti-inflammatory, and anti-apoptotic activity; however, during insulin resistance, the small HDL formation is promoted, producing a triglyceride-rich HDL3c subfraction, thus modifying its activity [63–65]. Complementarily, it must also be considered that Cd accumulation could be a decisive factor in metabolic toxicity. In this regard, the cadmium ion (Cd^{2+}) gets into the cells by different transporters such as the type 1 divalent metal transporter (DMT-1), zinc importer proteins (ZIP's), and voltage-gated calcium channels (VGCC) [3]. It has recently been demonstrated that not only does Cd^{2+} act as a substrate for the organic cation transporter 2 (OCT2) in a dose-dependent manner, but it also upregulates its expression and maximum transport rate (V_{max}), which might serve as a mechanism for Cd accumulation [10,66].

On the other hand, once Cd has caused a metabolic disruption, we began metformin treatment under the hypothesis that the drug is effective in metabolic control in a NOAEL dose (based on the mortality, biochemical, and body weight effects), regardless of the origin of the metabolic dysregulation, for example, hypercaloric diets consumption, energetic imbalances, neuroendocrinal disorders, and pharmacological and no-pharmacological supplementation, or in this case, by cadmium exposition. The metformin administration in rats without metabolic disorders does not show important biochemical changes in serum, but the drug positively modified the glycogen stored, as will be discussed later. In rats exposed to Cd and treated with a NOAEL dose of metformin, blood glucose levels are lowered and insulin resistance is ameliorated. Moreover, metformin elicits additional benefits, including improvement of lipid profiles, prevention of vascular complications, and lowering of the potential for hypoglycemia. Considering that metformin uses organic cation transporters (OCT1, OCT2, and OCT3) as an influx into the cells [29,30]. Rats that received one and two months of metformin treatment, including Cd exposure, showed a reduction in weight, abdominal perimeter, BMI, and percentage of body fat (Tables 2 and 3), which are results that were consistent with other works in both humans and Wistar rats that present metabolic complications by means other different to cadmium exposure [25,36]. These findings agree with the fact that metformin increases the AMP-activated protein kinase (AMPK), which is a "metabolic master switch"; its activation inhibits energy-consuming pathways and stimulates ATP-producing catabolic pathways [17]. In addition, AMPK activation can inhibit fatty acid synthesis by inhibition of acetyl-CoA carboxylase 1 and 2 and malonyl-CoA content reduction [17,67]. In this regard, a NOAEL dose of metformin is not completely effective because tissues showed a mild decrease in triglyceride storage after one month. However, in the second month, in tissues such as muscle, heart, and kidney, the triglyceride content increased (steatosis multi-tissue), even more than for the Cd group, although the liver observed a minimal recovery. The steatosis is a phenomenon common that is observed in an excess of metabolic needs or a limited rate of energy obtained via lipids (low lipid oxidation). Also, steatosis can generate lipotoxicity by lipotoxic intermediates, such as ceramide and acylcarnitine, and is promoted by triglyceride mobilization into VLDL (Figures 3 and 4), as shown by the lipid profile.

In relation to the serum lipid profile in rats exposed to Cd, after the treatment with metformin, a moderate improvement was observed in triglycerides, cholesterol, and the VLDL, LDL, and HDL fractions, but not at the level of the control group (Tables 2 and 3). Changes in cholesterol fractions also imply modifications of each subfraction (Figure 2). We observed that metformin had a limited action on lipid regulation because only the first month of treatment showed regulation, but in the second month, V1 and V2 subfractions followed the same pattern as the Cd group, suggesting a liver response to process lipidic clearance. However, almost all LDL subfractions were maintained in the control group, except for LDL-IVb, which is considered small, dense, and highly atherogenic [62]. It is interesting that the co-administration of cadmium and metformin treatment over-increases only this subfraction. The HDL showed a similar pattern because large and small subfractions were regulated, except for HDL3c, which is related to a poor glycemic control and an increase in the atherogenic risk [64]. The lipoprotein improvement is attributable to a lower insulin resistance and an enhancement of hepatic sensibility that is well recognized in metformin treatment [29,30,68,69]. However, in the presence of Cd, the time of metformin administration was probably insufficient because the hepatic sensibility was not completely restored, and the resistance was not eradicated, which explains the high FFA level coming from adipose tissue.

Additionally, a lower blood glucose was observed in both fasting and 1.75 g post-load of glucose/kg (Tables 2 and 3). Several mechanisms have been proposed to explain this action as a reduction of hepatic gluconeogenesis mediated by AMPK activation that inhibits the PKA pathway, diminishes the hepatic uptake of gluconeogenic substrates, and activates glycolysis [70–72]. At the muscle level, AMPK activation can also increase glucose consumption, optimizing energy expenditure and production, and participates in the transition glycolytic to oxidation of fatty acids [17]. In addition, the metformin treatment can improve the insulin action on the glycogenic pathway, as was observed in both groups, the metformin alone and Cd-metformin co-treated groups, in which glycogen concentration increased even more than the control group in almost all tissues analyzed [73,74]. Although the role of gluconeogenesis as a source of hepatic glucose overproduction and as a target of metformin action are well described, less is known about the role of changes in glycogen. However, our results suggest an increased activity of the glycogenic enzyme phosphoglucomutase and the decreased activity of glycogenolytic enzyme glycogen phosphorylase by the treatment, because glycogen phosphorylase is a rate-limiting enzyme of glycogenolysis and is regulated by phosphorylation and by the allosteric binding of AMP, ATP, glucose-6-phosphate, and glucose. Additionally, an effect observed in the group co-administered with Cd and metformin had a lactate increase (Tables 1 and 2). This was probably the result of an activity decrease of glucose 6-phosphatase and glucose 6-phosphate dehydrogenase, which had as a consequence an uncoupling of oxidative phosphorylation and the cycle of Krebs, which suppresses the lactate uptake, generating a high hepatic lactate production. This is a strong indicator of a low uptake of postprandial glucose [75–77]. The therapeutic doses of metformin usually cause little to no increase in basal and postprandial blood lactate levels (less than 1–2 mmol/L) but impair the hepatic metabolism. Also, a poor renal function by Cd accumulation associated with a metabolic kidney disruption could reduce lactate clearance [28].

Another important finding with the co-administration of Cd and metformin was hyperinsulinemia (Tables 1 and 2). It is recognized that metformin helps to restore the response to insulin, but not in insulin secretion [29,30]. However, the rats exposed to Cd develop an impaired insulinemic response. Some mechanisms for this have been proposed, such as insulin receptor impairment, low insulin receptor affinity by occupancy or negative cooperation, and a reduction in the number of receptors. Our results strongly suggest that Cd sensitize to the β-cells, producing a sustained hyperfunction and metformin would act on non-canonical pathways leading to the high insulin secretion [37]. It is probable that Cd exposure additional to metformin administration alters the ATP/ADP ratio that is permissive for K channel closure and enhanced insulin secretion being more susceptible in the presence of higher glucose concentration at the basal and postprandial [78]. An elevation in the total cellular NADH/NAD ratio also has been shown to promote insulin exocytosis [79]. The insulin release is correlated with an increase of intracellular Ca^{2+}, in which Cd^{+2} has been correlated. Changes in $\Delta\Psi_m$ are promoted by both glucose and metformin, which also met the set criteria as a potential factor important in insulin release [80].

In summary, is very important to consider the origin and duration of metabolic disruption for therapeutic management because Cd exposure has demonstrated metabolic toxicity in carbohydrates and lipids pathways, as well as serious alterations of insulin resistance in multiple tissues. Therefore, the treatment must consider intracell disorders, such as modifications in glycogen and triglycerides storage, as well as dysglycemia and dyslipidemia, particularly in subfractions of small LDL and HDL. In this sense, the treatment with a NOAEL dose of metformin in co-administration with Cd was limited with regard to metabolic regulation, and in the chronicity, which was counter-productive in relation to lipids storage in non-adipose tissue. However, the increase in dosage could bring unexpected consequences, such as morbidity, mortality, and clinical signs of toxicity, in addition to increasing metabolic acidosis (due to lactate and beta-hydroxybutyric acid). Although a NOAEL metformin dose was more effective in the carbohydrates' homeostasis, the associated metabolic pathways must be further studied and understood for establishing the therapeutic management in relation to the dosage and time of administration, selected on the basis of the metabolic disruption origin because the dose used in this work was demonstrated to be efficient in metabolic disorders from hypercaloric diet consumption, but not in cadmium exposition. Finally, the dosage selected must prevent clinical, metabolic, and toxicological complications since metformin is the first line of treatment for diabetes, obesity, overweight, insulin resistance, and other metabolic complications.

Author Contributions: V.E.S.-O., E.B., and S.T. designed the study and wrote the protocol. V.E.S.-O., A.D., J.Á.F.-H., U.P.-R., D.M.-G., V.A.-L., and S.T. performed the experiments. V.E.S.-O., E.B., and S.T. managed the literature searches and analysis, E.B., J.Á.F.-H., U.P.-R., and V.A.-L. undertook the statistical analysis. A.D., V.E.S.-O., E.B., and S.T. wrote the first draft of the manuscript. All contributing authors have approved the final manuscript.

Funding: This research received no external funding.

Acknowledgments: The authors thank Vicerrectoria de Investigación y Posgrado [VIEP; TRMS-NAT17-1] through Ygnacio Martinez Laguna, CONACyT and the "Sistema Nacional de Investigadores" of Mexico for the financial support of this research project [VESO, 533291]. We also thank Francisco Ramos Collazo [Bioterio Claude Bernard, BUAP] for his assistance and care of the animals used in this work and Thomas Edwards, Ph.D., for editing the English language text.

References

1. ATSDR, Agency for Toxic Substance and Disease Registry, U.S. Toxicological Profile for Cadmium. 2012. Available online: https://www.atsdr.cdc.gov/ToxProfiles/tp.asp?id=48&tid=15 (accessed on 15 May 2018).
2. Bernhoft, R.A. Cadmium Toxicity and Treatment. *Sci. World J.* **2013**, *2013*, 394652. [CrossRef] [PubMed]

3. He, L.; Wang, B.; Hay, E.B.; Nebert, D.W. Discovery of ZIP transporters that participate in cadmium damage to testis and kidney. *Toxicol. Appl. Pharmacol.* **2009**, *238*, 250–257. [CrossRef] [PubMed]

4. Nordberg, G.F. Historical perspectives on cadmium toxicology. *Toxicol. Appl. Pharmacol.* **2009**, *238*, 192–200. [CrossRef] [PubMed]

5. Zalups, R.K.; Ahmad, S. Molecular handling of cadmium in transporting epithelia. *Toxicol. Appl. Pharmacol.* **2003**, *186*, 163–188. [CrossRef]

6. Klaassen, D.C.; Liu, J.; Diwan, B.A. Metallothionein Protection of Cadmium Toxicity. *Toxicol. Appl. Pharmacol.* **2009**, *238*, 215–220. [CrossRef] [PubMed]

7. Habeebu, S.S.; Liu, J.; Liu, Y.; Klaassen, C.D. Metallothionein-null mice are more sensitive than wild-type mice to liver injury induced by repeated exposure to cadmium. *Toxicol. Sci.* **2000**, *55*, 223–232. [CrossRef] [PubMed]

8. Thevenod, F.; Ciarimboli, G.; Leistner, M.; Wolff, N.A.; Lee, W.K.; Schatz, I.; Keller, T.; AlMonajjed, R.; Gorboulev, V.; Koepsell, H. Substrate- and cell contact-dependent inhibitor affinity of human organic cation transporter 2: Studies with two classical organic cation substrates and the novel substrate Cd^{2+}. *Mol. Pharm.* **2013**, *10*, 3045–3056. [CrossRef] [PubMed]

9. Thevenod, F. Nephrotoxicity and the proximal tubule: Insights from Cadmium. *Nephron Physiol.* **2003**, *93*, 87–93. [CrossRef] [PubMed]

10. Thévenod, F.; Lee, W.K. Cadmium and cellular signaling cascades: Interactions between cell death and survival pathways. *Arch. Toxicol.* **2013**, *87*, 743–1786. [CrossRef] [PubMed]

11. Gobe, G.; Crane, D. Mitochondria, reactive oxygen species and cadmium toxicity in the kidney. *Toxicol. Lett.* **2010**, *198*, 49–55. [CrossRef] [PubMed]

12. Moulis, J.M. Cellular mechanisms of cadmium toxicity related to the homeostasis of essential metals. *Biometals* **2010**, *23*, 877–896. [CrossRef] [PubMed]

13. Moulis, J.M.; Thévenod, F. New perspectives in cadmium toxicity: An introduction. *Biometals* **2010**, *23*, 763–768. [CrossRef] [PubMed]

14. Takiguchi, M.; Yoshihara, S.I. New aspects of cadmium as an endocrine disruptor. *Environ. Sci.* **2005**, *13*, 107–116.

15. Silva, N.; Peiris-John, R.; Wickremasinghe, R.; Senanayake, H.; Sathiakumar, N. Cadmium a metalloestrogen: Are we convinced? *J. Appl. Toxicol.* **2012**, *35*, 318–332. [CrossRef] [PubMed]

16. Jiménez-Ortega, V.; Cano Barquilla, P.; Fernández-Mateos, P.; Cardinali, D.P.; Esquifino, A.I. Cadmium as an endocrine disruptor: Correlation with anterior pituitary redox and circadian clock mechanisms and prevention by melatonin. *Free Radic. Biol. Med.* **2012**, *53*, 2287–2297. [CrossRef] [PubMed]

17. McCarthy, A.D.; Cortizo, A.M.; Sedlinsky, C. Metformin revisited: Does this regulator of AMP-activated protein kinase secondarily affect bone metabolism and prevent diabetic osteopathy. *World J. Diabetes* **2016**, *7*, 22–33. [CrossRef] [PubMed]

18. Lee, B.K.; Kim, Y. Blood cadmium, mercury, and lead and metabolic syndrome in South Korea: 2005–2010. *Am. J. Ind. Med.* **2012**, *56*, 682–692. [CrossRef] [PubMed]

19. Satarug, S.; Moore, M.R. Emerging roles of cadmium and heme oxygenase in type-2 diabetes and cancer susceptibility. *Tohoku J. Exp. Med.* **2012**, *228*, 267–288. [CrossRef] [PubMed]

20. Chen, Y.W.; Yang, C.Y.; Huang, C.F.; Hung, D.Z.; Leung, Y.M.; Liu, S.H. Heavy metals, islet function, and diabetes development. *Islets* **2009**, *1*, 169–176. [CrossRef] [PubMed]

21. Hruby, A.; Hu, F.B. The epidemiology of obesity: A big picture. *Pharmacoeconomics* **2015**, *33*, 673–689. [CrossRef] [PubMed]

22. Mutlu, F.; Bener, A.; Eliyan, A.; Delghan, H.; Nofal, E.; Shalabi, L.; Wadi, N. Projection of Diabetes Burden through 2025 and Contributing Risk Factors of Changing Disease Prevalence: An Emerging Public Health Problem. *J. Diabetes Metab.* **2014**, *5*, 341.

23. Chen, L.; Magliano, D.J.; Zimmet, P.Z. The worldwide epidemiology of type 2 diabetes mellitus present and future perspectives. *Nat. Rev. Endocrinol.* **2011**, *8*, 228–236. [CrossRef] [PubMed]

24. Meigs, J.B. Epidemiology of type 2 diabetes and cardiovascular disease: Translation from population to prevention: The Kelly West award lecture 2009. *Diabetes Care* **2010**, *33*, 1865–1871. [CrossRef] [PubMed]

25. Treviño, S.; Sánchez-Lara, E.; Sarmiento-Ortega, V.E.; Sánchez-Lombardo, I.; Flores-Hernández, J.A.; Pérez-Benítez, A.; Brambila-Colombres, E.; González-Vergara, E. Hypoglycemic, lipid-lowering and metabolic regulation activities of metforminium decavanadate (H$_2$Metf)$_3$ [V$_{10}$O$_{28}$]·8H$_2$O using hypercaloric-induced carbohydrate and lipid deregulation in Wistar rats as a biological model. *J. Inorg. Biochem.* **2015**, *147*, 85–92. [CrossRef] [PubMed]

26. Desroches, S.; Lamarche, B. The evolving definitions and increasing prevalence of the metabolic syndrome. *Appl. Physiol. Nutr. Metab.* **2007**, *32*, 23–32. [CrossRef] [PubMed]

27. Cameron, A.J.; Shaw, J.E.; Zimmet, P.Z. The metabolic syndrome: Prevalence in worldwide populations. *Endocrinol. Metab. Clin. N. Am.* **2004**, *33*, 351–375. [CrossRef] [PubMed]

28. DeFronzo, R.; Fleming, G.A.; Chenc, K.; Bicsak, T.A. Metformin-associated lactic acidosis: Current perspectives on causes and risk. *Metab. Clin. Exp.* **2016**, *65*, 20–29. [CrossRef] [PubMed]

29. Viollet, B.; Foretz, M. Revisiting the mechanisms of metformin action in the liver. *Ann. Endocrinol. (Paris)* **2013**, *74*, 123–129. [CrossRef] [PubMed]

30. Viollet, B.; Guigas, B.; Sanz-Garcia, N.; Leclerc, J.; Foretz, M.; Andreelli, F. Cellular and molecular mechanisms of metformin: An overview. *Clin. Sci. (Lond.)* **2012**, *122*, 253–270. [CrossRef] [PubMed]

31. Summary of Revisions: Standards of Medical Care in Diabetes-2018. *Diabetes Care* **2018**, *4* (Suppl. 1), S4–S6. [CrossRef]

32. Chamberlain, J.J.; Herman, W.H.; Leal, S.; Rhinehart, A.S.; Shubrook, J.H.; Skolnik, N.; Kalyani, R.R. Pharmacologic Therapy for Type 2 Diabetes: Synopsis of the 2017 American Diabetes Association Standards of Medical Care in Diabetes. *Ann. Intern. Med.* **2017**, *166*, 572–578. [CrossRef] [PubMed]

33. Quaile, M.P.; Melich, D.H.; Jordan, H.L.; Nold, J.B.; Chism, J.P.; Polli, J.W.; Smith, G.A.; Rhodes, M.C. Toxicity and toxicokinetics of metformin in rats. *Toxicol. Appl. Pharmacol.* **2010**, *243*, 340–347. [CrossRef] [PubMed]

34. Zhou, G.; Myers, R.; Li, Y.; Chen, Y.; Shen, X.; Fenyk-Melody, J.; Wu, M.; Ventre, J.; Doebber, T.; Fujii, N.; et al. Role of AMP-activated protein kinase in mechanism of metformin action. *J. Clin. Investig.* **2001**, *108*, 1167–1174. [CrossRef] [PubMed]

35. Sanz, P. AMP-activated protein kinase: Structure and regulation. *Curr. Protein Pept. Sci.* **2008**, *9*, 478–492. [CrossRef] [PubMed]

36. Luong, D.Q.; Oster, R.; Ashraf, A.P. Metformin treatment improves weight and dyslipidemia in children with metabolic syndrome. *J. Pediatr. Endocrinol. Metab.* **2015**, *28*, 649–655. [CrossRef] [PubMed]

37. Treviño, S.; Waalkes, M.P.; Flores-Hernández, J.A.; León-Chavez, B.A.; Aguilar-Alonso, P.; Brambila, E. Chronic cadmium exposure in rats produces pancreatic impairment and insulin resistance in multiple peripheral tissues. *Arch. Biochem. Biophys.* **2015**, *583*, 27–35. [CrossRef] [PubMed]

38. Warnick, G.R.; Knopp, R.H.; Fitzpatrick, V.; Branson, L. Estimating low-density lipoprotein cholesterol by the Friedewald equation is adequate for classifying patients on the basis of the nationally recommended cut point. *Clin. Chem.* **1990**, *36*, 15–19. [PubMed]

39. Brunk, S.D.; Swanson, J.R. Colorimetric method for free fatty acids in serum validated by comparison with gas chromatography. *Clin. Chem.* **1981**, *27*, 924–926. [PubMed]

40. Rainwater, D.L.; Moore, P.H., Jr.; Gamboa, I.O. Improved method for making nondenaturing composite gradient gels for the electrophoretic separation of lipoproteins. *J. Lipid Res.* **2004**, *45*, 773–775. [CrossRef] [PubMed]

41. Bennett, L.W.; Keirs, R.W.; Peebles, E.D.; Gerard, P.D. Methodologies of tissue preservation and analysis of the glycogen content of the broiler chick liver. *Poult. Sci.* **2007**, *86*, 2653–2665. [CrossRef] [PubMed]

42. Nordberg, G.F.; Nogawa, K.; Nordberg, M.; Friberg, L. *Handbook of the Toxicology of Metals*, 4th ed.; Elsevier: Amsterdam, The Netherlands, 2014.

43. Haswell-Elkins, M.; Mcgrath, V.; Moore, M.; Satarug, S.; Walmby, M.; Ng, J. Exploring potential dietary contributions including traditional seafood and other determinants of urinary cadmium levels among indigenous women of a Torres Strait Island (Australia). *J. Expo. Sci. Environ. Epidemiol.* **2007**, *17*, 298–306. [CrossRef] [PubMed]

44. Skalnaya, M.G.; Tinkov, A.A.; Demidov, V.A.; Serebryansky, E.P.; Nikonorov, A.A.; Skalny, A.V. Hair toxic element content in adult men and women in relation to body mass index. *Biol. Trace Elem. Res.* **2014**, *161*, 13–19. [CrossRef] [PubMed]

45. Kawakami, T.; Nishiyama, K.; Tanaka, J.I.; Kadota, Y.; Sato, M.; Suzuki, S. Changes in macrophage migration and adipokine gene expression induced by cadmium in the white adipose tissue of metallothionein-null mice. *J. Toxicol. Sci.* **2012**, *37* (Suppl. II), AP-46.

46. Kawakami, T.; Sugimoto, H.; Furuichi, R.; Kadota, Y.; Inoue, M.; Setsu, K.; Suzuki, S.; Sato, M. Cadmium reduces adipocyte size and expression levels of adiponectin and Peg1/Mest in adipose tissue. *Toxicology* **2010**, *267*, 20–26. [CrossRef] [PubMed]

47. Schwartz, G.G.; Il'yasova, D.; Ivanova, A. Urinary cadmium, impaired fasting glucose, and diabetes in the NHANES III. *Diabetes Care* **2003**, *26*, 468–470. [CrossRef] [PubMed]

48. Flores, C.R.; Puga, M.P.; Wrobel, K.; Sevilla, M.E.G.; Wrobel, K. Trace elements status in diabetes mellitus type 2: Possible role of the interaction between molybdenum and copper in the progress of typical complications. *Diabetes Res. Clin. Pract.* **2011**, *91*, 333–341. [CrossRef] [PubMed]

49. Afridi, H.I.; Kazi, T.G.; Kazi, N.G.; Jamali, M.K.; Arain, M.B.; Sirajuddin Baig, J.A.; Kandhro, G.A.; Wadhwa, S.K.; Shah, A.Q. Evaluation of cadmium, lead, nickel and zinc status in biological samples of smokers and non-smokers hypertensive patients. *J. Hum. Hypertens.* **2010**, *24*, 34–43. [CrossRef] [PubMed]

50. Zhang, S.; Jin, Y.; Zeng, Z.; Liu, Z.; Fu, Z. Subchronic exposure of mice to cadmium perturbs their hepatic energy metabolism and gut microbiome. *Chem. Res. Toxicol.* **2015**, *28*, 2000–2009. [CrossRef] [PubMed]

51. Chapatwala, K.D.; Rajanna, B.; Desaiah, D. Cadmium-induced changes in gluconeogenic enzymes in rat kidney and liver. *Drug Chem. Toxicol.* **1980**, *3*, 407–420. [CrossRef] [PubMed]

52. Chapatwala, K.D.; Hobson, M.; Desaiah, D.; Rajanna, B. Effect of cadmium on hepatic and renal gluconeogenic enzymes in female rats. *Toxicol. Lett.* **1982**, *12*, 27–34. [CrossRef]

53. Han, J.C.; Park, S.Y.; Hah, B.G.; Choi, G.H.; Kim, Y.K.; Kwon, T.H.; Kim, E.K.; Lachaal, M.; Jung, C.Y.; Lee, W. Cadmium induces impaired glucose tolerance in the rat by down-regulating GLUT4 expression in adipocytes. *Arch. Biochem. Biophys.* **2003**, *413*, 213–220. [CrossRef]

54. Shanbaky, I.O.; Borowitz, J.L.; Kessler, W.V. Mechanisms of cadmium- and barium-induced adrenal catecholamine release. *Toxicol. Appl. Pharmacol.* **1978**, *44*, 99–105. [CrossRef]

55. Edwards, J.R.; Prozialeck, W.C. Cadmium, diabetes and chronic kidney disease. *Toxicol. Appl. Pharmacol.* **2009**, *238*, 289–293. [CrossRef] [PubMed]

56. Hectors, T.L.M.; Vanparys, C.; van der Ven, K.; Martens, G.A.; Jorens, P.G.; Van Gaal, L.F.; Covaci, A.; De Coen, W.; Blust, R. Environmental pollutants and type 2 diabetes: A review of mechanisms that can disrupt beta cell function. *Diabetologia* **2011**, *54*, 1273–1290. [CrossRef] [PubMed]

57. Larregle, E.V.; Varas, S.M.; Oliveros, L.B.; Martínez, L.D.; Anton, R.; Marchevsky, E.; Gimenez, M.S. Lipid metabolism in liver of rat exposed to cadmium. *Food Chem. Toxicol.* **2008**, *46*, 1786–1792. [CrossRef] [PubMed]

58. Fabbrini, E.; Magkos, F.; Mohammed, B.S.; Pietka, P.; Abumrad, N.A.; Patterson, B.W.; Okunade, A.; Klein, S. Intrahepatic fat, not visceral fat, is linked with metabolic complications of obesity. *Proc. Natl. Acad. Sci. USA* **2009**, *106*, 15430–15435. [CrossRef] [PubMed]

59. Sparks, J.D.; Sparks, C.E.; Adeli, K. Selective Hepatic Insulin Resistance, VLDL Overproduction, and Hypertriglyceridemia. *Arterioscler. Thromb. Vasc. Biol.* **2012**, *32*, 2104–2112. [CrossRef] [PubMed]

60. Messner, B.; Bernhard, D. Cadmium and cardiovascular diseases: Cell biology, pathophysiology, and epidemiological relevance. *Biometals* **2010**, *23*, 811–822. [CrossRef] [PubMed]

61. Messner, B.; Knoflach, M.; Seubert, A.; Ritsch, A.; Pfaller, K.; Henderson, B.; Shen, Y.H.; Zeller, I.; Willeit, J.; Laufer, G.; et al. Cadmium is a novel and independent risk factor for early atherosclerosis mechanisms and in vivo relevance. *Arterioscler. Thromb. Vasc. Biol.* **2009**, *29*, 1392–1398. [CrossRef] [PubMed]

62. Tellez-Plaza, M.; Jones, M.R.; Dominguez-Lucas, A.; Guallar, E.; Navas-Acien, A. Cadmium Exposure and Clinical Cardiovascular Disease: A Systematic Review. *Curr. Atheroscler. Rep.* **2013**, *15*, 356. [CrossRef] [PubMed]

63. Nakagawa, M.; Takamura, M.; Kojima, S. Some heavy metals affecting the lecithin-cholesterol acyltransferase reaction in human plasma. *J. Biochem.* **1977**, *81*, 1011–1016. [CrossRef] [PubMed]

64. Gomez-Rosso, L.; Lhomme, M.; Meroño, T.; Dellepiane, A.; Sorroche, P.; Hedjazi, L.; Zakiev, E.; Sukhorukov, V.; Orekhov, A.; Gasparri, J.; et al. Poor glycemic control in type 2 diabetes enhances functional and compositional alterations of small, dense HDL3c. *Biochim. Biophys. Acta* **2017**, *1862*, 188–195. [CrossRef] [PubMed]

65. Pari, L.; Ramakrishnan, L. Protective effect of Tetrahydrocurcumin on plasma lipids and lipoproteins in cadmium intoxicated rats. *Int. J. Toxicol. Appl. Pharmacol.* **2013**, *3*, 26–32.

66. Yang, H.; Guo, D.; Shu, Y. Cadmium Ion Upregulates the activity of Human Organic Transporter 2. *FASEB J.* **2017**, *31*, 819.10.

67. Fullerton, M.D.; Galic, S.; Marcinko, K.; Sikkema, S.; Pulinilkunnil, T. Single phosphorylation sites in Acc1 and Acc2 regulate lipid homeostasis and the insulin-sensitizing effects of metformin. *Nat. Med.* **2013**, *19*, 1649–1654. [CrossRef] [PubMed]

68. Yanovski, J.A.; Krakoff, J.; Salaita, C.G.; McDuffie, J.R.; Kozlosky, M.; Sebring, N.G.; Reynolds, J.C.; Brady, S.M.; Calis, K.A. Effects of Metformin on Body Weight and Body Composition in Obese Insulin-Resistant Children: A Randomized Clinical Trial. *Diabetes* **2011**, *60*, 477–485. [CrossRef] [PubMed]

69. Nasri, H.; Rafieian-Kopaei, M. Metformin: Current knowledge. *J. Res. Med. Sci.* **2014**, *19*, 658–664. [PubMed]

70. Foretz, M.; Hébrard, S.; Leclerc, J.; Zarrinpashneh, E.; Soty, M.; Mithieux, G.; Sakamoto, K.; Andreelli, F.; Viollet, B. Metformin inhibits hepatic gluconeogenesis in mice independently of the LKB1/AMPK pathway via a decrease in hepatic energy state. *J. Clin. Investig.* **2010**, *120*, 2355–2369. [CrossRef] [PubMed]

71. Mihaylova, M.M.; Shaw, R.J. The AMPK signaling pathway coordinates cell growth, autophagy and metabolism. *Nat. Cell Biol.* **2011**, *13*, 1016–1023. [CrossRef] [PubMed]

72. Madiraju, A.K.; Erion, D.M.; Rahimi, Y.; Zhang, X.M.; Braddock, D.T.; Albright, R.A.; Prigaro, B.J.; Wood, J.L.; Bhanot, S.; MacDonald, M.J.; et al. Metformin suppresses gluconeogenesis by inhibiting mitochondrial glycerophosphate dehydrogenase. *Nature* **2014**, *510*, 542–546. [CrossRef] [PubMed]

73. Lomako, J.; Lomako, W.M.; Whelan, W.G. The Biogenesis of Muscle Glycogen: Regulation of the Activity of the Autocatalytic Primer Protein. *BioFactors* **1990**, *2*, 251–254. [PubMed]

74. Vytla, V.S.; Ochs, R.S. Metformin Increases Mitochondrial Energy Formation in L6 Muscle Cell Cultures. *J. Biol. Chem.* **2013**, *288*, 20369–20377. [CrossRef] [PubMed]

75. DeFronzo, R.A. Pathogenesis of type 2 diabetes mellitus. *Med. Clin. N. Am.* **2004**, *88*, 787–835. [CrossRef] [PubMed]

76. Mithieux, G.; Guignot, L.; Bordet, J.C.; Wiernsperger, J.C. Intrahepatic Mechanisms Underlying the Effect of Metformin in Decreasing Basal Glucose Production in Rats Fed a High-Fat Diet. *Diabetes* **2002**, *51*, 139–143. [CrossRef] [PubMed]

77. Mithieux, G.; Rajas, F.; Zitoun, C. Glucose Utilization Is Suppressed in the Gut of Insulin-Resistant High Fat-Fed Rats and Is Restored by Metformin. *Biochem. Pharmacol.* **2006**, *72*, 198–203. [CrossRef] [PubMed]

78. Lamontagne, J.; Al-Mass, A.; Nolan, C.J.; Corkey, B.E.; Madiraju, S.R.M.; Joly, E.; Prentki, M. Identification of the signals for glucose-induced insulin secretion in INS1 (832/13) β-cells using metformin-induced metabolic deceleration as a model. *J. Biol. Chem.* **2017**, *292*, 19458–19468. [CrossRef] [PubMed]

79. Fu, A.; Robitaille, K.; Faubert, B.; Reeks, C.; Dai, X.-Q.; Hardy, A.B.; Sankar, K.S.; Ogrel, S.; Al-Dirbashi, O.Y.; Rocheleau, J.V.; et al. LKB1 couples glucose metabolism to insulin secretion in mice. *Diabetologia* **2015**, *58*, 1513–1522. [CrossRef] [PubMed]

80. Mugabo, Y.; Zhao, S.; Lamontagne, J.; Al-Mass, A.; Peyot, M.-L.; Corkey, B.E.; Joly, E.; Madiraju, S.R.M.; Prentki, M. Metabolic fate of glucose and candidate signaling and excess-fuel detoxification pathways in pancreatic β-cells. *J. Biol. Chem.* **2017**, *292*, 7407–7422. [CrossRef] [PubMed]

MDPI

St. Alban-Anlage 66

4052 Basel

Switzerland

Tel. +41 61 683 77 34

Fax +41 61 302 89 18

www.mdpi.com

Toxics Editorial Office

E-mail: toxics@mdpi.com

www.mdpi.com/journal/toxics

www.ingramcontent.com/pod-product-compliance
Lightning Source LLC
Chambersburg PA
CBHW051911210326
41597CB00033B/6111